Introduction to Applied Statistics

Introduction to Applied Statistics

A Modelling Approach

J. K. LINDSEY

Department of Social Sciences,
Univeristé de Liège
jlindsey@luc.ac.be

OXFORD
UNIVERSITY PRESS

This book has been printed digitally and produced in a standard specification
in order to ensure its continuing availability

OXFORD
UNIVERSITY PRESS

Great Clarendon Street, Oxford OX2 6DP

Oxford University Press is a department of the University of Oxford.
It furthers the University's objective of excellence in research, scholarship,
and education by publishing worldwide in

Oxford New York

Auckland Cape Town Dar es Salaam Hong Kong Karachi
Kuala Lumpur Madrid Melbourne Mexico City Nairobi
New Delhi Shanghai Taipei Toronto
With offices in
Argentina Austria Brazil Chile Czech Republic France Greece
Guatemala Hungary Italy Japan South Korea Poland Portugal
Singapore Switzerland Thailand Turkey Ukraine Vietnam

Oxford is a registered trade mark of Oxford University Press
in the UK and in certain other countries

Published in the United States
by Oxford University Press Inc., New York

© J. K. Lindsey, 1995, 2004

The moral rights of the author have been asserted

Database right Oxford University Press (maker)

Reprinted 2009

ISBN 978-0-19-852895-1

Preface to the first edition

To consult the statistician after an experiment is finished is often merely to ask him to conduct a *post mortem* examination. He can perhaps say what the experiment died of. To utilise this kind of experience [the research worker] must be induced to use his imagination, and to foresee in advance the difficulties and uncertainties with which, if they are not foreseen, his investigation will be beset. (R.A. Fisher, 1938, *Sankhyā* 4, 14–17)

This preface is primarily addressed to the person who will be responsible for teaching a course based on the book. For the student who will be following such a course, the final chapter (7) is meant for you. You might like to try reading it now, and compare your reactions with those when you reread it at the end of studying the material in this book. Of course, you can also read this preface now to try to get some idea of what your instructor is up to!

In modern applied statistics, the analysis of categorical and duration data plays a central role. Data following a normal distribution are relatively rare in practice in most scientific disciplines, so that this distribution serves primarily as an approximation when nothing better is available. Unfortunately, introductory text books in statistics do not reflect these elementary facts. Nonstatistics majors see only means, variances, 'least squares' regression and analysis of variance, with the accompanying tests, and, if lucky, a little about frequency tables. No wonder that they come away from such courses believing that statistics is a difficult subject of little practical use. Such is the heritage of the precomputer era.

The present text is based on the contents of an introductory statistics course that I have taught to social science majors for almost twenty years, first at the University of British Columbia, Vancouver, in the mid seventies, and then at the University of Liège, Belgium. There, the students subsequently have two more advanced courses using GLIM (Lindsey, 1989 and 1992).

The material covered here should be appropriate to introduce the fundamentals of statistics as a first course in any nonmathematical discipline, such as medicine, biology, psychology, education, sociology, or geography. The examples and exercises have been chosen to cover such areas. Models for time-dependent data are included in a final chapter, because of their importance in many fields, such as economics. However, other specialised areas have had to be neglected in order to cover adequately the base material for a firm understanding of statistics. Economists may feel the need to add supplementary material on linear normal models, psychologists on experimental design and multivariate normal models, sociologists on sample surveys, medical workers on testing, and so on.

I have also long held the secret belief that the approach presented here should be used as a first introduction to statistics even for the most mathematically sophisticated students. The feel for the aims of statistics should be clearly communicated before elaborate mathematical justifications are presented. However, I have not yet had the opportunity to test this hypothesis by introducing mathematics students to statistics in this way.

Although the course is designed for the computer age, it can also be taught with only the use of a hand calculator (which is what I do). This is most appropriate if a second course, using a computer, and more specialised to the students' discipline, is to follow. The idea is that it is essential that the beginning student see exactly how the basic results are obtained, instead of relying on the free ride provided by 'black box' computer software. However, if the student is likely only to have this one course, it will be preferable to do at least part of the examples and exercises using a computer.

The course is built around the construction of models to describe the structure of data. The basic building block is the probability distribution. A model describes how such a distribution changes form in different subgroups of a population. As well, the form of the distribution, itself, may indicate how the data were generated. Thus, inference, from a sample to the population, plays a secondary role; the text centres on developing an understanding of how to interpret a likelihood function. Minimal mathematical knowledge is required: primarily manipulation of logarithms, of variables with indices, and of elementary functions.

The use of rather mathematical notation, such as subscripts, sums (Σ), and logarithms, may seem beyond the average nonmathematics student. After all, they are supposed to detest mathematics, are they not? However, a bit of time spent near the beginning of the course, showing, for example, how such a sum is actually concretely done using a calculator, will bring light to their eyes as they realise that they too can do higher mathematics, and that it is useful for them! In my experience, any student who is capable of success at university can master the material in this book, if properly taught and motivated.

For the course to be successful, the students must spend a large part of the time actually trying to understand and analyse data, applying models to them. For the examples and exercises, I have tried to supply the maximum available information on each data set. Unfortunately, this has been a difficult task. Most scientific journals do not allow the publication of raw data, while statisticians, and their journals, are not generally interested in how data were generated. They look for nice illustrations of some procedure they have discovered and are not concerned with 'Material and Methods'. (As I complete this manuscript, I receive a letter from a major 'applied' statistics journal rejecting a paper, one main reason being that it contains too much biological detail.) Students must become as familiar as possible with the ways in which data are actually generated. But, at the same time, they must learn how statistical models provide enough abstraction to be applicable in vastly different subject areas.

On the other hand, students must not be allowed to become lost in the details of the calculations; interpretation of the results obtained is the essential part of the learning process. One fruitful approach is to administer a short questionnaire to all students near the beginning of the year, with a few suitably chosen questions. These, then, can be used as examples to illustrate concretely the usefulness of the various models. The application of a logistic model, in Chapter 2, to a simple 2×2 frequency table obtained from such data can tell the students things they did not imagine about their class!

The material covered here has been used for a two semester course of about 60 hours of lectures, plus practical tutorial work. This may seem like a slim volume for such a substantial course, but it is essential that the students spend a lot of time both seeing demonstrations of fitting a large number of models to data and being guided in doing so themselves. In the practical sessions, they should be fitting the models to real data using a pocket calculator or computer. Data tables are supplied in the exercises, but, where possible, these should be replaced with material directly relating to other courses that the students are concurrently taking. In this way, they will have background information about where the data come from and what questions can be asked of them, things that cannot be supplied in a series of textbook exercises. Nevertheless, this text could be presented in considerably less lecture time, but with the attendant loss of practical experience with data analysis.

For those wishing to go to more advanced modelling after the completion of this material, the book by Dobson (1990) can be recommended as an excellent starting point. For categorical data models, look at Agresti (1990), Collett (1991), or Lindsey (1995). The approach to inference used here is developed in more detail in Kalbfleisch (1985).

The tables in the Appendix were created with GLIM4. The graphics were drawn with MultiPlot, for which I thank Alan Baxter. I would also like to thank all of the contributors of data sets; they are individually cited when each table is first presented.

Thanks for detailed comments and discussion of this material go to Murray Aitkin, Rob Crouchley, Richard Davies, Bruno Genicot, Gentiane Haesbroeck, Dan Heitjan, Philippe Lambert, Patrick Lindsey, and Joe Whittaker. The helpful comments and suggestions of the seven OUP referees were also greatly appreciated. Without the untiring understanding, help, and competence of the Oxford University Press editorial staff, especially Elizabeth Johnston and Anna Drage, this work would not have been possible. Obviously, the lively discussion with students taking the course over the past two decades and pointing out the unclearness of my ways was also indispensable.

Diepenbeek and Liège   J.K.L.
August, 1994

Preface to the second edition

In the six years since I completed the first edition of this text, the role of statistical modelling has grown in applied statistics. *Ad hoc* tests are being replaced by interval estimation of parameters, often including study of the shape of likelihood functions. The increased emphasis on parameter estimates necessarily leads to more careful consideration of the models involved.

Unfortunately, the way that most students are introduced to statistics has not fundamentally changed. Few receive instruction in the encompassing and unifying role of models in extracting systematic structure from the randomness in empirical data. Even as innovative a text as Gelman and Nolan (2002) is 'a bag of tricks'.

Approaches to statistics can be classified in many ways. Two are useful here:

- the goal of a statistical analysis may be
    (1) to advance human knowledge by obtaining previously *unknown* results (scientific inference) or
    (2) to aid in making a personal or collective choice among some finite number of *known* possibilities (decision-making).
- the variability studied may be assumed to arise primarily from
    (1) the natural states of the *phenomena* under study (likelihood approach),
    (2) the ways in which the *data* are collected (frequentist approach), or
    (3) the lack of complete knowledge of the phenomena in the *mind* of the person conducting the analysis (Bayesian approach).

The approach taken in this text emphasises the first category of both classifications. However, this does not imply that models are not useful, and even essential, to the other categories as well.

Most of the addition pages in this new edition provide expanded details about the material already present in the first edition. I have tried to give more explanations where relationships among concepts were previously often implicit. A lot of the material has been reordered. Many elementary concepts previously scattered through the text have been brought forward to Chapter 1. Chapter 4 is thoroughly reorganised. I have also tried to separate more clearly the examples from the more theoretical parts so that the instructor can replace the former by ones more appropriate to the interests of a particular class. Little really new material has been included. I have, however, added four new distributions in Chapter 4, the beta-binomial, Laplace, Cauchy, and Student t.

I have retained a presentation that allows all material to be taught only with the aid of a hand calculator (something which I continue to do). However, I also supply, on my web site www.luc.ac.be/~jlindsey, all of the data sets and the R code used in the analysis of the examples. As well, the graphs in the text have all be redrawn using R. An instructor's manual is also available on the same site.

Bruno Genicot carefully read the whole manuscript and Patrick Lindsey several chapters; both provided many useful comments and suggestions. A referee made a number of suggestions that improved several sections of the text.

Liège   J.K.L.
May, 2003

Contents

1
Basic concepts

The role of scientists is not simply to try to describe phenomena, but rather to *explain* them. From available knowledge, they develop scientific theories of why something happens and then make appropriate observations to check them. In turn, these observations can lead to modifications of the theory, which must also be checked, and so on. Scientific research involves a continual interaction between theory and empirical observation. Thus, a theory is constantly being checked, corrected, and improved in the light of new data; when this is no longer possible, it is replaced, but only if a better one is available.

Most scientific theories can be formulated as *models*, simplified explanations of reality. These allow the essential characteristics of the phenomenon to be highlighted and better understood. An essential characteristic of empirical data collected for such scientific purposes is *variability*. Any useful model must take this into account. Thus, no two human beings are identical; even a pair of measurements in physics will generally not be absolutely identical. Statistics is the fundamental tool in the study of this variability.

In this text, I shall be primarily interested in introducing the areas of statistics that are useful in scientific research about living beings, and especially that involving human beings. The methodology is essentially the same for research on all living organisms, but some extra problems arise in dealing with human subjects.

We should first notice that, in any given study, with its specific goals, all types of observable variability will not be of equal interest.

Examples
In a *sample* of people, the responses to each given question will be different depending on the person answering, that is, they will vary across the sample. However, the importance placed on these differences for any specific question will depend on the goals of the study. Consider a question on smoking in two different research projects.

- In the study of lung cancer, any *systematic* differences in the responses between smokers and nonsmokers may be important, whereas *variability* among people within either of these categories, such as nationality might not be. The latter would simply be considered to be due to chance or *ran-*

dom, because it is not pertinent to the goal of the study.
- In a study of the education levels of people, one might be looking for *systematic* differences among nationalities, and any differences due to smoking could be ignored, as *random*.

□

Our first goal must be to formalise these vague ideas—sample, variability, systematic, random—in order to be able to conduct a study and subsequently to extract the pertinent information from the data that have been collected. Any such process necessarily involves rather major approximations to reality. As we have just seen, in any empirical study, we can treat variability in at least two quite distinct ways, somehow separating the systematic from the random differences. Thus, by developing the distinction between these two types of differences in data, I shall be able to formalise the idea of variability.

We shall learn how this process can yield an approximation to reality known as a *statistical model*. The development and application of such models in the context of empirical data is the subject of this text. In this way, statistics reflects the scientific endeavour, with interaction between theory and data: the theory (the model) can tell us what kind of data are required and the data may indicate how to modify the theory, that is, the statistical model.

1.1   Variables

The first step must be to begin to clarify what we shall actually observe or measure. Useful information can occur in many contexts. It may be obtained at first hand, such as by measuring the height of each person, or at second hand, by asking the people how tall they are. But what forms can this information take? For statistical analysis to be possible, we must be very careful about how the information is recorded.

1.1.1   DEFINITION

Generally, in scientific research, we hope that our theory will apply in a very wide context. Almost invariably, the group or *population* involved in applying the theory will be so large that all members cannot be directly observed. Otherwise, the scientific interest of the theory will be too limited.

Suppose, then, that we select some relatively small group of individuals, called a *sample*, from that population. (In Section 1.4, I shall discuss some appropriate ways in which such a group can be chosen.) These may be people, but might also be any entity of interest, such as hospitals, classrooms, families, or villages. This will be our *unit of observation*. The choice of an appropriate unit can be very important.

Example
Suppose that we want to study the numbers of children in the families in our pop-

ulation of interest. If we question a sample of children (the unit of observation) to obtain information about their brothers and sisters, we shall certainly have biased results: we shall have no families without children. (What other biases will be present?) Thus, the unit of observation must be the family and this must be clearly defined. If a couple is separated and not remarried, does this count as one or two families?   □

Variable construction   We can now divide the observed group up according to differences in the characteristic of interest, which every individual must have. This might, for example, be their sex, the colour of their hair, or their opinion on some subject. In the example above, it is the number of children. The individuals are classified into different categories according to the characteristic that each has. In order for this classification to be useful, each individual must have only one type of the characteristic and all individuals must be included. In other words, the classification must involve categories that are

- *mutually exclusive* and
- *exhaustive* (or all inclusive).

I shall call such a classification a *variable* because the characteristic varies across the individual units of observation.

I shall represent such a variable abstractly by some capital letter, say Y. The label on each type or category is a *value* of the variable. Then, I shall take the corresponding small letter, here y, to represent such a value. Thus, the possible values for the variable sex would be male and female, for hair colour they might be brown, black, blond, ... Because there must be more than one value, I shall distinguish among them by an index: y_1, y_2, and so on.

Possible problems   We have now seen that a variable must have values that are exhaustive and mutually exclusive. Certain problems can be encountered in the construction of variables. These include

(1) *mixed criteria of classification*: the variable values, French, German, Flemish, Basque, Muslim, or Jewish mix citizenship, nationality, and religion;

(2) *multiple entries*: a questionnaire allows respondents to choose more than one answer, such as several sports played, so that the number of replies will be greater than the number of individuals in the sample;

(3) *overlapping categories*: the father of a child is a manual worker, a farmer, or a factory employee, subgroups that are not mutually exclusive.

If a questionnaire is used to obtain the data, each question may not correspond to one variable, although such a correspondence is usually preferable. Sometimes, however, the reply to a given question necessarily involves separate parts that must be handled as distinct variables.

Students sometimes confuse the concept of variable with its values, say speaking of the variable male. However, male cannot *vary*; it is a specific fixed *value* of the variable sex.

1.1.2 CHARACTERISTICS OF OBSERVATIONS

Types of variables   Not all variables are the same. They may be classified in various ways according to the characteristics of their values.

(1) All values of variables, even numbers, are names. However, if these names have no relationship of order or magnitude among them, the variable is called *nominal*. Sex, religion, and nationality are of this type.

(2) If a variable can have three or more distinct values and these have a rank order, but no measure of distance among them, the variable is *ordinal*. Examples include letter grades for an examination, A, B, C, D, E, and the appreciation of some object as like, indifferent, dislike.

(3) If the values are *counts*, the variable is *integral*. Thus, as above, we might have the number of children in each family, the individual unit of observation being the family.

(4) If the values are measures, the variable is called *continuous*. Here, we have age, income, and so on. Continuous variables may be divided into two types.

    (a) *Ratio* variables have a natural zero point, so that division makes sense.

    (b) *Interval* variables have no natural zero. Saying that one temperature is twice another is meaningless: the answer is different on the Celsius, Kelvin, and Fahrenheit scales.

All variables that are not continuous are called *discrete*. Integral and continuous variables are *quantitative*, whereas nominal and ordinal variables are *qualitative* or *categorical*. The values of a qualitative variable are sometimes called *attributes*. A variable that has only two possible values is called *binary* or *dichotomous*, whereas any other categorical variable is *polytomous*.

To a certain extent, the position of any variable on this list is arbitrary. Variables can be moved up the list but not down. Thus, age might be taken as integral (number of years), ordinal (one person older than another), or nominal, but distinct religions could not be counts.

Roles of variables   The values of a variable classify the members of a group into a number of categories. This is true whether the variable is qualitative or quantitative. Even for a continuous variable, such categories must always exist because any measuring instrument can only have finite precision. We never measure the age of human beings very precisely, certainly not even to the nearest second! Thus, for such variables, we always have some *unit of measurement*, such as months or years. Notice that this may not be the unit in which measurements are expressed; a length of 3.14 metres has a unit of measurement of centimetres, not metres.

If the classification specified by a variable is so detailed that every individual has a different value, all individuals are unique. Without strong, unverifiable assumptions, general conclusions cannot be drawn about the group as a whole (except that everyone is different). The variable is of no interest, statistically speaking, and probably not scientifically either. All variation has been made systematic; none is random. Instead, we shall be interested in relations among groups, within which the individuals cannot be distinguished. Within a group, all individuals are considered to be *exchangeable* with respect to other characteristics that are not considered to be relevant for the current question of interest.

Although statistics involves the observation of individuals, their individuality or uniqueness is immediately lost with the collection of the data and their classification according to the values of a variable. This is a requirement for statistical modelling to be possible. But, for most studies of human beings, there is a second reason: the participants must be allowed to remain anonymous.

Accuracy and precision When we observe a variable, the *accuracy* refers to how close we come to the 'true' value. This applies to both qualitative and quantitative variables.

Examples
A thermometer that consistently measures 2° too hot is not accurate. If people misunderstand a question or lie about their age, the answers will not be accurate.□

On the other hand, the *precision* refers to how finely we can observe the values of a variable.

Example
Age measured to the nearest month is more precise than that to the nearest year.□

Both accuracy and precision must be taken into account in preparing any study.

Example
A questionnaire that asks for income to the nearest Euro may provide very precise results but, if many respondents lie, the conclusions will not be very accurate! On the other hand, if people are asked to situate their incomes within a series of categories each 500 Euros wide, they may be less inclined to lie so that the results may be more accurate, but at the cost of being less precise. □

In conducting a study, one may have to play off accuracy and precision against each other in order to obtain the most useful results.

1.1.3   SEVERAL VARIABLES

In most situations, more than one variable will be observed on each individual. Among them, two basic kinds of variables can be distinguished.

Response variable   Most often, one variable (or more) will be the scientific focus of the study. This is called the *response variable*. In the construction of statistical models, it is the variable that will be assumed to involve randomness in the study, in the sense that has been used intuitively above and will be defined more strictly in Section 1.3 below. I shall reserve Y to represent it abstractly, with y_1, y_2, and so on, corresponding to specific observed values.

Example

If Y represents the counts of the numbers of children in families, then y_1 for a specific family might equal two children.   □

One should not, however, assume that the same variable will be chosen as response for all analyses of a data set. The choice may change with the problems to be studied and the questions to be answered.

Explanatory variables   Other variables can serve to divide the population into subgroups, for example by sex. Often, when we construct models, these variables are assumed not to contain any randomness, but simply to have fixed values. I shall call each of them an *explanatory variable*. Another name often used is *covariate*. I shall represent such a variable abstractly by X, and its observed values by x_1, x_2, and so on. Usually, more than one explanatory variable will be available. Then, I shall distinguish among these by an index: $X_1, X_2, \ldots$. In such a case, the observed values will require two indices: $x_{11}, x_{12}, \ldots$ for the first explanatory variable, $x_{21}, x_{22}, \ldots$, for the second, and so on.

In each of the subpopulations so defined, it will be possible that the way in which the response variable Y varies is different. Then, the response is said to *depend* on that explanatory variable defining the subpopulations.

Example

The proportion of people replying correctly to a question (the response variable) may change with age (the explanatory variable).   □

Both response and explanatory variables can be of any of the types described in Section 1.1.2.

1.2   Summarising data

We now know in what form our data must be (variables), but we do not yet know how to collect them in a representative and objective way suitable to our goals (Section 1.4). Nevertheless, let us now look at some ways in which empirical data can be summarised. This can depend on the type of variable involved.

Table 1.1. A data table.

Individual	Sex	Age	Opinion	$\cdots$
1	M	25	For	$\cdots$
2	M	34	For	$\cdots$
3	F	22	Against	$\cdots$
4	M	18	Indifferent	$\cdots$
$\vdots$	$\vdots$	$\vdots$	$\vdots$	$\ddots$

Tables and graphics are two important forms of *descriptive statistics*. These will be useful for checking for errors, for learning what basic information is contained in the data, and for presentation to other people. Such descriptive techniques contrast with models that generally are for *explanation*. However, these techniques can also be useful both in understanding models and in performing simple visual checks on them.

1.2.1 TABLES

At the same time that one constructs a variable that one plans to observe, one should also decide exactly how it will be recorded.

Raw data The fundamental table of any study will contain the observed values of all the variables, the raw data from the sample of individuals. Such a table will usually take a standard form. Each column will correspond to a different variable, with the first often being the identification number of the individual. Each line will contain the information from a different individual. In this way, we obtain a data matrix such as that in Table 1.1.

This information will be stored electronically in a data file for use in a computer. The first few lines of the file may contain information describing exactly what data are contained in it. The line immediately preceding the actual data should contain the names of the variables; they will be at the top of each column. All subsequent analysis will be derived from this table, so that it is essential that it does not contain errors. This table will *not* generally be shown to people (outside the research team) interested in learning about the results of the study!

Once the raw data table is available, ways of summarising the information in it will be necessary.

Frequencies One of the fundamental ways to summarise raw data is to calculate the number of individuals observed to be in each category of a variable, called the *frequencies*. These will be integer numbers so that they may appear similar to counts, mentioned above as one type of variable. However, I shall use the term frequencies to refer to aggregations of *different* individuals and counts to refer to a characteristic of one individual unit of observation.

Example
In a sample, there might be 53 families with 1 child and 37 families with 2 children. Then, 1 and 2 children are counts, whereas 53 and 37 families are the corresponding frequencies of families having these characteristics, where the family is the unit of observation (the 'individual').   □

Because a count is a value of a variable, if it is say a response variable, it would be represented by Y with values $y_1, y_2, \ldots$. In contrast, I shall use n_i to represent the frequency for the category i of any variable (not just counts) with I distinct categories in all. In the following text, I shall often use letters as lower indices, such as this i, to indicate abstractly to which subgroup I am referring.

Then, the total size of the sample will be given by summation, represented by

$$n_1 + n_2 + \cdots + n_I = \sum_i n_i$$
$$= n_\bullet$$

Both $\sum$ and the dot in place of the index are abbreviations to indicate that a sum is being taken. Depending on the situation, we may find one more convenient than the other.

One variable   A *frequency table* gives the number of individuals observed to be in each category of a variable. It can be constructed by enumerating the total number of individuals in each category of the variable, perhaps with suitable aggregation of adjacent categories. Such results may also be presented as relative frequencies, obtained by dividing the frequencies by the total number $n_\bullet$ (usually represented by N in the legend of the table), or as percentages, obtained by multiplying the relative frequencies by 100.

The values of the variable, or the interval limits, will be listed, usually vertically on the left-hand side and the (relative) frequencies or percentages will be placed beside them. The legend should clearly describe the variable and how the data were obtained. The total number of observations should be stated. If there are missing responses, not used in the table, their number should also be indicated.

Example
The frequency distribution for the sizes of families in the USA in March 1966, expressed in percentages, is given in Table 1.2. Of course, such a table, as it stands, raises questions about how the variable, family size, was constructed. What is the definition of family? Why are there no families of size one? How were the families selected? Was it a sample or the whole population? As well, the total number is not given!   □

For quantitative variables, the finest classification is by the unit of measurement used in recording the raw data. However, although these raw data should always be used for analyses, for summaries it is often useful to have a coarser

Table 1.2. Distribution of families in the USA in March 1966, by their size. (Reproduced from Mueller *et al.*, 1970, p. 48, from *Current Population Reports*, 1967, USA Government Printing Office)

Family size (persons)	Per cent
2	33.6
3	20.2
4	19.3
5	12.8
6	7.1
7 or more	7.0
Total	100.0

grouping. The loss of information is compensated by the greater ease of understanding. Thus, we can choose the number of *class intervals* and the *width* of each. The intervals are *closed* if two limits are specified. On the other hand, one or both the end intervals (containing the extreme values) may be *open*, with only one limit fixed.

Example
The raw data may contain the exact ages of people in years. However, a frequency table may be more easily understood if these are grouped into closed categories of say five years: 16–20, 21–25, and so on. If there are few old people, it may be useful to create one large open category, say 61 years or older.　　　　□

Several variables Simple frequency tables do not show the relationships among variables. To do this, the data may be displayed in a more complex cross-classified frequency or *contingency table*. In the simplest case, such a table will show the relationship between two variables. The right-hand column and the last line of the table usually contain the sums of the row and column frequencies respectively. They are called the *marginal totals*, because they are on the margins of the table. In fact, they are just the separate frequencies for each of the variables, as would be found in the simple frequency tables for one variable, described above (if there are no missing values of one or the other variable).

Example
With two variables, a two-way table will show the frequencies for each combination of values of the variables, X and Y. Table 1.3 provides the $n_\bullet = 28$ observations cross-classifying sex (X) and post-graduate plans (Y) for the first students to take this course, in 1976.　　　　□

It is now time to introduce some further notation. We may represent the frequencies in a two-way table more abstractly by n_{ij}. Again, i will indicate a

Table 1.3. Graduate study plans of the first students taking this course, in 1976.

Sex	PhD plans		Total
	Yes	No	
Male	5	12	17
Female	6	5	11
Total	11	17	28

Table 1.4. Notation for frequencies in two-way tables.

X	Y				Total
	y_1	y_2	$\cdots$	y_I	
x_1	n_{11}	n_{21}	$\cdots$	n_{I1}	$n_{\bullet 1}$
x_2	n_{12}	n_{22}	$\cdots$	n_{I2}	$n_{\bullet 2}$
$\vdots$	$\vdots$	$\vdots$	$\ddots$	$\vdots$	$\vdots$
x_J	n_{1J}	n_{2J}	$\cdots$	n_{IJ}	$n_{\bullet J}$
Total	$n_{1\bullet}$	$n_{2\bullet}$	$\cdots$	$n_{I\bullet}$	$n_{\bullet\bullet}$

specific category of one of the variables, in this text generally the response Y and j the other (this may not be true in other books and scientific papers). The total numbers of categories will be, respectively, I and J.

The marginal totals are obtained by summation:

$$n_{i1} + n_{i2} + \cdots + n_{iJ} = \sum_j n_{ij}$$
$$= n_{i\bullet}$$
$$n_{1j} + n_{2j} + \cdots + n_{Ij} = \sum_i n_{ij}$$
$$= n_{\bullet j}$$

In a similar way, the overall total is

$$\sum_i \sum_j n_{ij} = n_{\bullet\bullet}$$

Again, I have used both the sum $\sum$ and the dot notation. Here, they indicate addition of all values, indexed by j, in a given column, i, and vice versa for rows. This notation is illustrated in Table 1.4.

Example (continued)
In Table 1.3, $n_{11} = 5$, $n_{12} = 6$, $n_{21} = 12$, and $n_{22} = 5$. The marginal totals are $n_{1\bullet} = 11$, $n_{2\bullet} = 17$, $n_{\bullet 1} = 17$, and $n_{\bullet 2} = 11$. The overall total is $n_{\bullet\bullet} = 28$. □

Relative frequency or percentage tables may also be constructed. Usually, they will be calculated for the response variable, within each subgroup of the

Table 1.5. The cross-classified percentages for Table 1.3.

Sex	PhD plans Yes	No	Total	N
Male	29	71	100	17
Female	55	45	100	11
Total	39	61	100	28

Table 1.6. Summary of the notation using the data on graduate study plans of the first students taking this course, in 1976.

Sex	PhD plans Yes	No	Total
Male	$n_{11} = 5$ (29%)	$n_{21} = 12$ (71%)	$n_{\bullet 1} = 17$ (100%)
Female	$n_{12} = 6$ (55%)	$n_{22} = 5$ (45%)	$n_{\bullet 2} = 11$ (100%)
Total	$n_{1\bullet} = 11$ (39%)	$n_{2\bullet} = 17$ (61%)	$n_{\bullet\bullet} = 28$ (100%)

explanatory variable(s). In such cases, the direction in which the calculations have been made should always be indicated, for example by providing a row or column of say 100%.

Example (continued)
For Table 1.3, the percentages are given in Table 1.5. From this table, we see more clearly that the proportion of students with graduate plans, corresponding to the response Y, is quite different in the two subgroups of sex, defined by the explanatory variable X. Having graduate plans appears to depend on sex. At the same time as clarifying this relationship, the table provides sufficient information to reconstruct the original data.

For reference, a more complete presentation, summarising what has preceded, is shown in Table 1.6. However, such a table is far too overloaded to be used for presentation to the general reader. □

Cross-classified contingency tables involving more than two variables are more complex, but follow the same principles.

1.2.2 MEASURING SIZE AND VARIABILITY

When a variable is quantitative, several special descriptive statistics are available. Usually, they are of most interest for response variables. Let y_i represent a distinct value of the response variable, with n_i observations of that value. If all observed values of the response variable are different, all $n_i = 1$. Then, the following quantities can always be calculated from the data.

(1) The *mode* is that value of y_i with the highest frequency, that is, the largest n_i. It is the only one of these quantities that can be used with nominal variables. (Why?)

(2) The *median* is that value of y_i that separates the observations into two equal parts, one half of them being smaller and the other half larger. It is the only one of these quantities that is not changed if the variable is transformed, for example by taking logarithms (see Section 4.6).

(3) The *estimated* or *empirical mean* is

$$\bar{y}_\bullet = \frac{1}{n_\bullet} \sum_i n_i y_i \tag{1.1}$$

Here, the bar on $y_\bullet$ indicates an average.

(4) The *minimum* and *maximum* values of y_i.

(5) The *estimated* or *empirical variance* is

$$s^2 = \frac{1}{n_\bullet} \sum_i n_i (y_i - \bar{y}_\bullet)^2$$
$$= \frac{1}{n_\bullet} \sum_i n_i y_i^2 - \bar{y}_\bullet^2$$

Often, the *standard deviation*, s, is also used. Notice that the variance only measures *symmetric* variability about the mean.

The first three quantities provide indications of the size of the responses, or of their *location* on the horizontal axis of a histogram (Section 1.2.3). The other two indicate the variability or *dispersion* of the responses. In particular, the variance gives the dispersion about the mean, that is, of the width of the histogram. (Another possibility would be to use the sum of absolute values of differences; see Section 4.4.3.) The mean and the variance are called the first two *moments* for a quantitative variable.

Calculating variances It will be useful to know how to manipulate variances because they provide an approximate measure of variability in many circumstances. Two rules are:

(1) The (estimated) variance of the sum or difference of responses is the sum of their (estimated) variances.

(2) If all responses are multiplied by a constant, the (estimated) variance is multiplied by the square of that constant.

Notice that the variance of subtracted values is a sum not a difference!

Example
Suppose that we have $n_\bullet$ responses with estimated variance s^2. Then, the estimated variance of the sum of these responses will be $n_\bullet s^2$, that of the difference of two responses, say $y_1 - y_2$, will be $2s^2$, and that of the mean of the responses

$$\frac{n_\bullet s^2}{n_\bullet^2} = \frac{s^2}{n_\bullet}$$

In this last calculation, the sum of the responses is multiplied by $1/n_\bullet$ to obtain the mean and this multiplicative constant has to be squared. □

The 'standard deviation' of any quantity calculated from the raw data has a special name: the *standard error*. Thus, the standard deviation describes an observable characteristic (a variable) of the population, whereas the standard error describes some calculated quantity (such as the mean) that we cannot observe directly. The two should not be confused.

Example
The estimated mean is a quantity calculated from the raw data. Its estimated variance is $s^2/n_\bullet$ so that its estimated standard error is $s/\sqrt{n_\bullet}$. Because the estimated variance and standard deviation are characteristics of the population, their value would not be expected to change a great deal as the sample size increases, but would instead become stable. In contrast, we see that the estimated standard error will decrease rather quickly with larger samples. In such samples, we know more precisely what the value of the mean is.

Because the estimated variance and standard deviation are calculated from the raw data, they will also each have a variance and a standard error! □

1.2.3 GRAPHICS

The most efficient way to present summaries of data is often by graphical methods. If well done, these can be understood easily and quickly by the reader. Modern computer software can produce many apparently beautiful plots, but one must be wary that they actually do clearly convey the desired results. Two types of plots are most useful in the context of statistical modelling: histograms and scatter plots.

Histograms A *histogram* is a graphical representation of a variable constructed from a (relative) frequency or percentage table. It is most useful to display the variability in a response variable.

In simple cases, the values of the variable are usually given horizontally and the frequencies or percentages vertically. If the variable is quantitative and has no values near zero, the horizontal axis need not start at zero. However, the vertical axis must begin at zero.

Example
The family size data from Table 1.2 are plotted in Figure 1.1. Obviously, the way in which the first or last category is represented when the interval is open is rather arbitrary, as for the 7+ category in this example. □

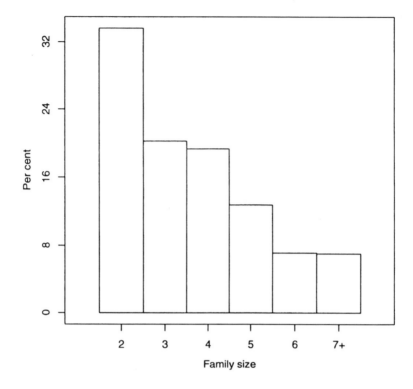

Fig. 1.1. Histogram of the distribution of families by size.

If the variable is categorical, a histogram is often called a bar chart. Here, such a distinction is not useful because we wish to treat all (response) variables in essentially the same way. Statistical modelling provides a unified approach to all types of variables.

If the values of a variable are strictly nominal, the categories have no particular order and the shape of the histogram can be arbitrarily changed by reordering. An order that is appropriate for the message to be conveyed should be chosen.

Example
Suppose that a variable has six nominal values, labelled A, B, C, D, E, and F, where the letters do not indicate any ordering. The corresponding observed frequencies are 9, 7, 10, 2, 5, and 7. Four histograms, among the 720 possible, are shown in Figure 1.2.   □

For continuous variables, or when an integral variable is grouped, the standard way to construct the histogram is to let the *area* of each bar, and not its height,

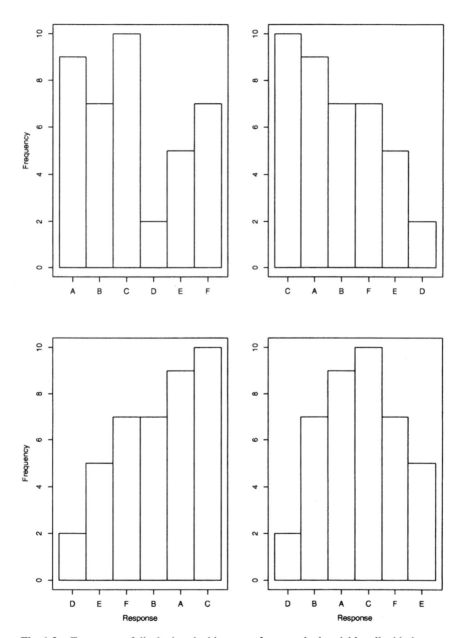

Fig. 1.2. Four ways of displaying the histogram for a nominal variable, all with the same frequencies.

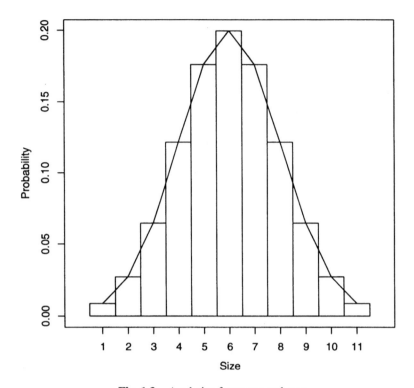

Fig. 1.3. A relative frequency polygon.

represent the frequency. When the values are quantitative, they obviously cannot be reordered arbitrarily as can nominal variables. The form of the histogram for quantitative variables is very important, providing useful information about the data, as we shall see in Chapter 4. Often, it can be visualised more clearly by joining the midpoints of the top of each bar to create a *frequency* or *probability polygon*, as in Figure 1.3.

Although individual histograms are primarily useful for illustrating the variability in a response variable, explanatory variables can also be integrated into the display by using a set of histograms. A separate histogram can be produced for each value of the explanatory variable. However, care must be taken that the scales of the axes on all the histograms are identical so that comparisons can be made.

Example (continued)
For the students' PhD plans in Table 1.3, the corresponding histograms are plotted in Figure 1.4. Again, we see how their plans appear to differ with (depend on) sex.   □

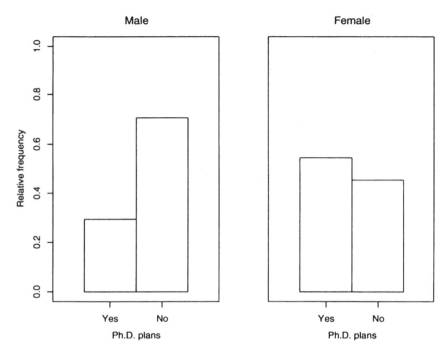

Fig. 1.4. Proportions of students of the two sexes planning to do a PhD, from Table 1.3.

Table 1.7. Weights (kg) of people before and after a diet. (Dobson, 1990, p. 24)

Before	64	71	64	69	76	53	52	72	79	68
After	61	72	63	67	72	49	54	72	74	66

Both frequency tables and histograms can be used to summarise the information in the data. The values of the variable, defined by the unit of measurement, may sometimes be combined into categories with larger class intervals. This evidently involves loss of information, but can be valuable for highlighting the essential characteristics of the variable.

Scatter plots   If several of the variables are quantitative, they may be plotted graphically, in pairs, as points in a *scatter plot*. Usually, the response variable is placed on the vertical axis.

Example
Data on change in weight with diet are given in Table 1.7. The weights are of ten people, before and after going on a high carbohydrate diet for three months. The response variable Y is the weight after diet, whereas the explanatory variable X is the weight before diet. These are plotted in Figure 1.5. As might be

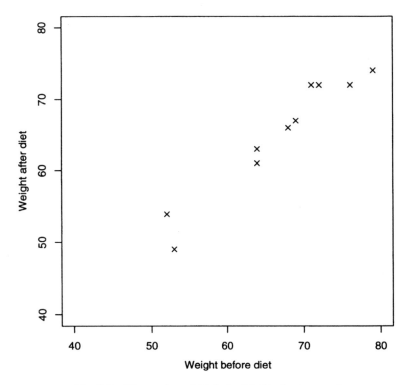

Fig. 1.5. Change in weight (kg) with diet for ten people.

expected, weight after diet is rather closely related to weight before diet, as seen
more clearly in the scatter plot than in the table.   □

Simultaneous representation of the relationships among more than two variables is difficult on a flat surface. When only three variables are to be represented,
contour or 3-D plots may be used. In more complex situations, combinations of
such graphs may be useful. For example, a collection of scatter plots can be used,
but these will only show the pairwise relationships among variables. I shall not
discuss these methods here.

Other possibilities   Many other ways of representing data graphically are available, for example, by box-and-whisker plots, dot charts, stem-and-leaf plots, and
so on. In the context of modelling, none of these add substantial information
over those presented above. If interested, the reader is referred to a good book on
descriptive statistics or statistical graphics for further details.

However, some methods should definitely not be used. Thus, one common
graph found in nonscientific publications is the pie chart. This is extremely dif-

ficult to judge accurately because human beings cannot easily compare areas of different sizes. *Pie charts should be avoided.* Instead use histograms because these involve linear comparisons.

1.2.4 DETECTING POSSIBLE DEPENDENCIES

Tables and plots involving two or more variables can often be used to discern relationships among the variables.

Example (continued)
In both Table 1.5 and Figure 1.4 above, we saw that the proportion of students with graduate plans was quite different in the two subgroups defined by their sex. The possibility of having graduate plans appeared to depend on sex. □

Similar conclusions may often be obtained from a scatter plot. If there are enough points on the graph, those close to an imaginary vertical line, placed at some given value of the explanatory variable, may provide an idea of the relative frequencies of values of the response for that value of the explanatory variable. This can be imagined as a histogram lying along that line.

Now suppose that, as the explanatory variable is changed, that is, as the imaginary line is moved to the left or right, the set of points moves up or down. Then, the relative frequencies of the responses are changing with, and thus depend on, that variable. In other words, the histogram is changing form and position as that line is moved. (See also Figure 5.1 for a special example.)

Example (continued)
Although there are not enough points in Figure 1.5 really to discern the form of the relative frequencies of the response, weight after diet, we can see that it is indeed moving up as the explanatory variable, weight before diet, increases (as the imaginary vertical line moves from left to right). □

Thus, one of our fundamental goals will be to develop models to describe more formally, and hopefully help to explain, how the relative frequencies of the responses change in, or depend on, the different subgroups of a sample. I shall do this in Chapters 2 and 5. Then, one must be able to draw conclusions about what this means more generally for the population of interest (Chapter 3). However, before attempting any such complex analysis of a data set, one should study closely tables and graphs. If nothing else, this permits one to detect erroneously recorded values and to obtain a 'feel' for the information contained in the data.

1.3 Probability

Up until now, I have been primarily concerned with the data that will be collected. It is now time to lay the basis for model construction. Here, I shall describe the basic building blocks. Putting them together actually to construct a realistic model must await Chapter 2.

1.3.1   DEFINITION

Many terms are in everyday use to describe variability and the unknownness associated with it: probability, random, chance, likelihood, plausibility, and so on. However, in order to have scientifically useful tools, any such concepts must be given precise and restrictive definitions. Here, I shall examine closely how *probability* and *random* can be defined; in Chapter 3, I shall provide the definition of *likelihood*. Thus, from now on, it is wise to reserve these three terms for use only with their strict statistical meaning. For example, 'probable' and 'likely' will refer to very different things.

Populations and samples   Any group of individuals that we wish to study will be called the *population*. It must be clearly specified in such a way that we know which individuals belong to it or, at least, so that we know if any given individual belongs to it. However, often it will be so large that we cannot possibly observe all the individuals in it, whether because of time constraints, expense, or other reasons. Then, a *sample* is any subgroup of the population that we choose to observe.

Usually, we want the sample to be 'representative' of the population; I shall describe how this can be accomplished in Section 1.4.2. We must assume that the individuals are exchangeable, as far as possible, for all their specific characteristics that do not interest us. Most often, we shall also require that the observations selected from the population be *independent*: observing one tells us nothing about which others may be selected. I shall give a more rigorous definition of independence shortly.

Data generating mechanisms   Now let us look more closely at what happens when we classify individuals belonging to some group of interest, the population, according to the values of some response variable. The various subgroups or subpopulations, labelled by the values of the variable, will have differing numbers of individuals in them. This has come about through some mechanism that has distributed the individuals among the subgroups. I shall call this the *data generating mechanism*. Statistics, in its applications in science, is primarily concerned with studying such mechanisms. A *statistical model* is an abstract construction to describe, and help to understand, a data generating mechanism, as a more or less useful approximation to that reality.

Examples
At conception, human beings are distributed into two subgroups by sex. The data generating mechanism, creating this two-category distribution, primarily revolves around the genetic lottery. A statistical model will provide a simple means by which individuals end up with one sex or the other (Section 4.3.3).

The process by which an individual develops an opinion for or against abortion is quite different, although it also only involves two categories. Nevertheless, it might possibly be represented by a similar statistical model to that just hinted at for sex.   □

Frequencies and probabilities   Consider first a population of individuals where every member is observed. Corresponding to each distinct value of a variable, say y_i, there will be a calculable *frequency*, say n_i, that is the number of individuals in the subpopulation i. This set of numbers forms the *frequency distribution* of the variable for that particular population. If, instead, we look at the corresponding *proportions* of individuals in each subpopulation, this is called the *relative frequency distribution* of the variable for the given population.

Example (continued)
Suppose that the population in which we are interested has $n_{\bullet} = 28$ individuals, as in Table 1.3, and that the variable is sex, with subpopulations of 17 males and 11 females. The frequency distribution is $n_1 = 17$, $n_2 = 11$, where the index 1 refers to males and 2 to females. From these, the proportions of males and females in the population can be calculated: $17/28 = 0.61$ and $11/28 = 0.39$, the relative frequency distribution.   □

If an individual is selected from the population by chance, so that all individuals have an equal possibility of being chosen, the selection is said to be completely *random*.

Example (continued)
When one individual is chosen from our student population at random, there are 17 chances in 28 of picking a male and 11 chances in 28 of picking a female. Notice that we are assuming that all males are exchangeable, as are all females. For our present purposes, all other characteristics are ignored. The choice of any given male in the population signifies the same thing to us as if any other male had been chosen, and the same for the females.   □

The *probability* that an individual, chosen at random, has a given value of the variable is the relative frequency of that value in the population. Thus, with random selection, individuals have equal probability of being chosen, but values of a variable generally do not (unless there are equal numbers of the values in the whole population). We may write

$$\Pr(\text{observe individual } i) = \frac{1}{N}$$

where N is the number of individuals in the population, but

$$\text{Pr(observe the given category } i) = \text{Pr(observe the given value } y_i)$$
$$= \text{Pr}(Y = y_i)$$
$$= \pi_i$$
$$= \frac{n_i}{N}$$

Notice that this definition of probability does not depend on the need to repeat the selection more than once. It is a theoretical construct describing the population.

Example (continued)
For our data, the probability that one individual chosen at random is male is $\text{Pr(male)} = \pi_1 = 0.61$ and female is $\text{Pr(female)} = \pi_2 = 0.39$. □

For small populations, which can be completely observed, the relative frequency distribution, in our example, π_1, π_2, is also the *probability distribution*. However, most often the population of interest will be so large that the probability distribution cannot be studied directly, as it has been in our example. Only a sample from the population (see Section 1.4) can be observed. The population and a sample from it are distinct entities. Then, I shall distinguish between

- the population with its probability distribution, a theoretical construction which is generally unobservable, and
- a sample from it, with its relative frequency distribution that can actually be calculated.

When the observed group is the whole population, it is called a *census* instead of a sample.

Thus, from now on, the relative frequency distribution will refer only to the sample. I shall represent relative frequencies by $\hat{\pi}_i$ to indicate that they are calculated from only part of the population, the sample. These are *estimates* of the corresponding population values, not necessarily equal to those for the whole population. In our example above, the two were identical. However, in general, the relative frequencies are always calculable (once a sample is observed), whereas probabilities usually are not. Then, given the representativity of the sample, we shall hope that the estimates can tell us something useful about the corresponding overall population (Chapter 3).

Properties of probabilities   Any set of I numbers with the following two properties may be called a probability distribution:
  (1) Each number of the set must lie between zero and one. (In fact, it is sufficient that they all be non-negative.)
  (2) The sum of all the numbers in the set must be unity. (If it is not, the numbers can be 'normalised' by dividing each one by their total.)

Written mathematically, we have

$$0 \leq \pi_i \leq 1 \qquad \text{for all } i$$

$$\pi_1 + \pi_2 + \cdots + \pi_I = \sum_i \pi_i$$

$$= 1$$

We can see that, if, and only if, a variable is properly constructed, as described in Section 1.1.1, so that its values are exhaustive and mutually exclusive, shall we obtain a probability distribution.

In fact, this definition can be slightly extended. The number of different values I of a variable may be allowed to be infinite if the sum of the corresponding probabilities can still be calculated and equals one (Section 4.1.2). This can only occur if most of the infinite number of values of π_i are extremely small. Such an extension can be useful for deriving some mathematical results, but of course an infinite number of different values of a variable can never actually be observed. The theoretical construction takes this into account in that the probabilities of observing most of the values must be extremely small.

Parameters A probability distribution refers to the proportions of individuals with the various values of the variable in the population. It is a set of unknown and unobservable quantities, called *parameters*, represented by the π_i. These are the basic building blocks of statistical models. Such models will apply to some population of interest, information about them being obtained by observing a sample from that population.

Note that confusion is possible because, in many scientific fields, the term parameter refers to observable, often controllable, characteristics of the objects under study. These are what statisticians call variables. In our terminology, parameters can never be directly observed. Indeed, because the population cannot usually be completely observed, such parameters cannot even be exactly calculated.

Extensions Once the concept of probability is understood as it refers to the subgroups of a population, it can be extended to apply in other contexts. These will often be processes, usually evolving over time (sometimes in space), where a clear definition of a population is not easy.

Example
Suppose that the variable of interest is the number of unemployed in a given country. This can vary from zero to the total working population. Moreover, it will change over time, recorded perhaps every month. Some complex data generating mechanism is producing the series of values of the variable, the numbers of unemployed, and statistical models can be constructed to attempt to understand it. However, there is no clear 'population' of all 'individuals' that might be observed. All months? For how long? □

Thus, the definition of probability given above can be extended to apply to any phenomenon whose outcome is not predetermined before it is observed. If it is suspected that some data generating mechanism of interest is at work, and a suitable variable can be constructed, various possible statistical models can usually be found to help to try to understand the phenomenon. Although most of this text is concerned with models applied to concrete populations, I shall look at a number of examples of processes in Chapter 6.

1.3.2 PROBABILITY LAWS

We now need to look a little more rigorously at a few basic ideas about probability. Let us call an *event* something that happens, or has happened, to an individual. In a sample, the event is observed, and recorded, as the value of a variable. A *simple event* is the smallest unit of observation; a *compound event* is composed of several simple events occurring together.

Example
A simple event could be that a human being was born either male or female. A compound event would be that the individual is female and aged 29.   □

As we have seen, the probability of an event y_i is defined as the frequency of the event in the population, say n_i, divided by the total number of events, $\sum_i n_i = N$, in that population:

$$\Pr(y_i) = \frac{n_i}{N}$$

(In this section, for clarity, I shall use $\Pr(y_i)$ instead of π_i to indicate the probability of the event y_i.) Thus, if all events have equal probability, with I different possible events, the probability of each is

$$\Pr(y_i) = \frac{1}{I} \quad \text{for all } i$$

Joint, conditional, and marginal probabilities   The *joint probability* of two (or more) events, that is, of a compound event, is the probability that they occur together: $\Pr(y_1 \text{ and } x_1)$. In such a situation, $\Pr(y_1)$ and $\Pr(x_1)$ are called the *marginal probabilities*. Notice that I use values of two different variables because, by our construction of variables, two values of the same variable cannot occur simultaneously for the same individual.

The *conditional probability* that an event y_1 occurs given that we know that another event x_1 has already occurred is represented as $\Pr(y_1|x_1)$. Obviously, if x_1 has occurred, $\Pr(x_1) > 0$. Joint, marginal, and conditional probabilities are then related by the equation

$$\Pr(y_1|x_1) = \frac{\Pr(y_1 \text{ and } x_1)}{\Pr(x_1)} \tag{1.2}$$

This may also be rewritten

$$\Pr(y_1 \text{ and } x_1) = \Pr(y_1|x_1)\Pr(x_1) \tag{1.3}$$

so that the joint probability of two (or more) events can always be decomposed as a product.

As my notation indicates, a conditional probability is usually used to refer to a response variable, the condition being specified by an explanatory variable.

Example
The conditional probability of an individual being a smoker may be different depending on the sex: $\Pr(\text{smoker}|\text{male}) \neq \Pr(\text{smoker}|\text{female})$. □

We shall soon see that conditional probabilities will form the basis of our statistical models. They can be used to describe how *random* variability can differ *systematically* in various subpopulations. In this way, we have a more formal means of describing dependence between two variables.

Example (continued)
From either Table 1.5 or Figure 1.4 above, the probability of a student planning to do a PhD is different for the two sexes. In this very small population, $\Pr(\text{PhD}|\text{male}) = 0.29$ and $\Pr(\text{PhD}|\text{female}) = 0.55$. □

By symmetry, Equation (1.3) yields

Bayes' formula

$$\Pr(y_1|x_1)\Pr(x_1) = \Pr(x_1|y_1)\Pr(y_1) \tag{1.4}$$

□

Independence   If two events are *independent*, then I have said above that observing one tells us nothing about the other. We are now in a position to formalise this idea.

Example
Suppose that a survey of some opinion is conducted orally in class. Each student is asked in turn to give his or her opinion. Then, the answers given by the last students questioned may be heavily influenced by those they have heard from others in the class. The opinions expressed probably will not be independent. □

More precisely, if y_1 and x_1 are independent, the probability of y_1 does not change, whether we know x_1 or not (and vice versa). This can be written succinctly as

$$\Pr(y_1|x_1) = \Pr(y_1) \qquad (1.5)$$

But, from Equation (1.2), we immediately see that

$$\Pr(y_1|x_1) = \frac{\Pr(y_1 \text{ and } x_1)}{\Pr(x_1)}$$
$$= \Pr(y_1)$$

or

$$\Pr(y_1 \text{ and } x_1) = \Pr(y_1)\Pr(x_1) \qquad (1.6)$$

Thus, we have

The product law
The probability of a compound event is the product of the probabilities of the simple events of which it is composed, if the simple events are independent of each other.   □

Example
Suppose that two (exchangeable) people are asked independently (not orally in the presence of each other!) a question for which the probability of a correct answer is $\frac{1}{6}$. Then, the probability of both giving the correct answer is

$$\Pr(\text{two correct answers}) = \frac{1}{6} \times \frac{1}{6}$$
$$= \frac{1}{36}$$

I shall elaborate on and nuance this result in Section 1.3.4.   □

Equations (1.5) and (1.6) are strictly equivalent, but the meaning of the first is more intuitively obvious in the context of scientific modelling.

Independence may refer to events on the same or on different individuals. On the same individual, it may refer to repeatable events from the same variable or to different variables. When events from different variables are independent, we say that the variables are independent. Two events cannot be both exhaustive and independent, nor can they be both mutually exclusive and independent.

Example
A survey questionnaire asks the sex and an opinion of each respondent in a sample. If individuals are chosen at random, the sexes of any two people in the sample should be independent. If the opinion does not depend on the sex of the individual, those two variables, sex and opinion, are said to be independent. If the survey

is repeated six months later with the same people, the opinion of an individual
this second time will usually not be independent of that given the first time.   □

We now have a means of detecting independence. At least in a sample, we
can check whether or not the equality in Equation (1.5) or (1.6) holds.

Example (continued)
From Table 1.3 above, the marginal probability of a student planning to do a PhD
is $\Pr(\text{PhD}) = 0.39$. This is not equal to the conditional probabilities, $\Pr(\text{PhD}|\text{male}) =$
0.29 and $\Pr(\text{PhD}|\text{female}) = 0.55$.

Similarly, the marginal probability of being male is $\Pr(\text{male}) = 0.61$. The
product of the marginal probabilities is $\Pr(\text{PhD})\Pr(\text{male}) = 0.39 \times 0.61 = 0.24$.
This is not equal to the joint probability, $\Pr(\text{PhD and male}) = 0.18$.

Unfortunately, comparison of these numbers can give quite different indications of the extent of dependence: 0.39 with 0.29 or 0.55, 0.29 with 0.55, 0.24
with 0.18.   □

Thus, if there is dependence, we still have no way of measuring how strong it
is. The development of such measures will have to wait until Section 2.1.

Alternative events Consider now alternative events, where one simple event *or*
another can occur. The most common examples will correspond to two different
values of the same variable. In such cases, when the events are mutually exclusive,
the total probability is the sum of the probabilities of the simple events:

$$\Pr(y_1 \text{ or } y_2) = \Pr(y_1) + \Pr(y_2)$$

This is

The addition law
The probability of one out of two (or more) mutually exclusive events occurring
is the sum of the individual probabilities.   □

Examples
If a question has four possible answers, A, B, C, or D, and someone replies randomly, with equal chance of choosing any answer, the probability of replying A
or C is

$$\Pr(\text{reply A or C}) = \frac{1}{4} + \frac{1}{4}$$
$$= \frac{1}{2}$$

Consider also the total area of several bars in the histogram in Figure 1.3
above.   □

If the two events are not mutually exclusive, the calculation is more complex. This will generally occur for values of different variables. Both events may, then, occur together. So as not to count such possibilities twice, we must subtract their joint probability:

$$\Pr(y_1 \text{ or } x_1) = \Pr(y_1) + \Pr(x_1) - \Pr(y_1 \text{ and } x_1)$$

Obviously, if the events are mutually exclusive, they cannot occur together and $\Pr(y_1 \text{ and } x_1) = 0$. Recall also that, if the two events are independent, then $\Pr(y_1 \text{ and } x_1) = \Pr(y_1)\Pr(x_1)$ may be substituted into this formula.

The product and addition laws are basic properties of probability from which many other useful results can be derived.

Example
Suppose that y_1 and y_2 are mutually exclusive and exhaustive events, so that $\Pr(y_1 \text{ and } y_2) = 0$ and $\Pr(y_1 \text{ or } y_2) = 1$. For any other event, x_1, we have

$$\begin{aligned}\Pr(x_1) &= \Pr(x_1 \text{ and } y_1) + \Pr(x_1 \text{ and } y_2) \\ &= \Pr(x_1|y_1)\Pr(y_1) + \Pr(x_1|y_2)\Pr(y_2)\end{aligned}$$

so that the probability of *any* event can be decomposed into a sum using conditional probabilities.   □

Another important application will be given in Section 1.3.4 below.

Expected value   In Section 1.2.2, we saw how to calculate the mean and variance of a sample when the response is quantitative. At least theoretically, we can also calculate them for the whole population of size N.

The theoretical mean μ_T is defined in the same way as the empirical mean in Equation (1.1), but applies to the whole population:

$$\begin{aligned}\mu_T &= \sum_i \frac{n_i y_i}{N} \\ &= \sum_i \pi_i y_i \\ &= \mathrm{E}(Y)\end{aligned}$$

Because $\pi_i = n_i/N$ in the population, this mean is also defined as the sum of all possible responses multiplied by their corresponding probabilities. The theoretical mean is also given a very confusing name: the *expected value* of the response Y, hence the symbol $\mathrm{E}(Y)$. In almost all cases, no one would ever expect the response variable to have the value given by $\mathrm{E}(Y)$.

Examples
Suppose that the response is the number of children in each family. Then, its expected value will almost certainly not be an integer.

Now consider a binary response variable with possible values zero and one and corresponding probabilities π_1 and π_2. The theoretical mean or expected value is

$$0 \times \pi_1 + 1 \times \pi_2 = \pi_2$$

In both cases, the expected value is an impossible value!   □

As we have already seen, the estimated value of the mean from a sample is $\widehat{\mu_T} = \bar{y}_\bullet$.

The expected value can also be found for functions of a response variable. One of special interest is $(Y - \mu_T)^2$:

$$\begin{aligned}
\mathrm{E}[(Y - \mu_T)^2] &= \sum_i \pi_i (y_i - \mu_T)^2 \\
&= \sum_i \frac{n_i (y_i - \mu_T)^2}{N} \\
&= \sigma_T^2
\end{aligned}$$

This is the theoretical variance for the whole population. Its estimated value from a sample is $\hat{\sigma}_T^2 = s^2$.

If, for mathematical convenience, we assume that there is an infinite number of possible values of the response, such as all the non-negative integers, it may still be possible to calculate expected values. If the response variable is continuous, the sum must be replaced by an integral. In both cases, if most of the probabilities are small enough, the expected value can be calculated; otherwise, it will be infinite. Thus, the response variable for a theoretical infinite population may have an infinite mean and variance, although any existing finite population, or sample from it, cannot.

1.3.3 PLOTTING PROBABILITIES

The procedure for plotting a probability distribution as a histogram is identical if the variable is qualitative or if the intervals of a quantitative variable are all of unit width. As we saw in Section 1.2.3, if the interval widths chosen are different from one, the procedure is slightly more complex, because the area, not the height, is important.

By convention, the total area in the graph of a probability distribution is set equal to one, the sum of all the probabilities. The area π_i of bar i in the histogram is then given by multiplying its height by its width. The result must equal the probability of that value of the variable:

$$\pi_i = f_i \times \Delta_i \tag{1.7}$$

where f_i is the height and Δ_i the width. Because the probability and the width are known when creating such a graph, we can calculate the required height:

$$f_i = \frac{\pi_i}{\Delta_i}$$

The height and the probability are the same only if the width is one. The heights f_i are known as the *probability densities*.

By the law of addition, the area under the curve between any two points on the horizontal axis of a probability histogram or polygon, is equal to the probability of the value of the variable lying between those two values.

Example
In Figure 1.3, the probability of a size between 4.5 and 7.5 is about 0.48.   □

Histograms are constructed for empirical data for which only a finite number of categories can be observed. However, theoretical probabilities may involve an infinite number of categories (Section 4.1.2), in which case the areas indicating the probabilities of the observed categories may sum to less than one.

1.3.4   MULTINOMIAL DISTRIBUTION

We are now familiar with the meaning of the probability of observing a value of a variable, what I have also called the probability of a simple event. However, we are most often interested in studying the 'simultaneous' observation of a number of individuals, a sample from the population. This is an important special example of a compound event. Here, I shall apply some of the above results on probability laws to obtain one of the most fundamental probability distributions in statistical modelling.

Suppose that we have a sample of $n_\bullet$ independent observations for a variable with I different possible values. These will be recorded in some order, say

$$y_3, y_1, y_I, \ldots, y_2$$

with corresponding probabilities

$$\pi_3, \pi_1, \pi_I, \ldots, \pi_2$$

(Here, for simplicity, I revert to using π_i for the probability of an event, as in Sections 1.2 and 1.3.1.) However, there will usually be more than one observation for at least some of the values; these are the frequencies n_i. Because the observations are independent, we can multiply their probabilities together:

$$\pi_3 \times \pi_1 \times \pi_I \times \cdots \times \pi_2 = \pi_1^{n_1} \pi_2^{n_2} \cdots \pi_I^{n_I}$$
$$= \prod_i \pi_i^{n_i}$$

Note that, in any particular sample, some of the n_i may possibly be zero and others one. Notice also how I use $\prod$ as an abbreviation for multiplication, similar to $\sum$ for addition.

We have here the probability of observing the given sample, *in one given order*, if the units were distinguishable. However, we assume that our individuals are exchangeable so that they are not distinguishable, except by the variable observed. In other words, our given set of observations could also have occurred in a number of other different, but equivalent, orders. Because of the independence and exchangeability, the order is immaterial.

Example
If we have five independent observations of a question with yes or no as answer and obtain two yes answers, these could have been observed in the order YYNNN, YNYNN, or YNNYN, and so on. Because of the assumption of independence, the order in which the sample is obtained will not alter our conclusions. □

The total number of such orders or *combinations* is given by the *combinatorial*

$$\frac{n_\bullet!}{n_1!n_2!\cdots n_I!} = \binom{n_\bullet}{n_1\ n_2\cdots n_I}$$

where $n!$, called n factorial, is defined as

$$n! = n \times (n-1) \times (n-2) \times \cdots \times 2 \times 1$$

Thus, because these orderings are mutually exclusive possibilities for any one sample of size $n_\bullet$, we can add their probabilities, each order having the same, to calculate the total probability for a given sample:

$$\Pr(n_1, n_2, \ldots, n_I) = \binom{n_\bullet}{n_1\ n_2\cdots n_I} \prod_i \pi_i^{n_i} \tag{1.8}$$

This now refers to observation of the individuals in any order. It is called the *multinomial distribution*. The parameters of this distribution are the set, $\pi_1, \ldots, \pi_I$. They determine the form of the distribution. Because they sum to unity, only $I-1$ are really necessary to know this form.

Binomial distribution An important special case of the multinomial distribution occurs when the variable has only two distinct values:

$$\Pr(n_1, n_2) = \binom{n_\bullet}{n_1} \pi_1^{n_1} (1 - \pi_1)^{n_2} \tag{1.9}$$

where $\pi_2 = 1 - \pi_1$. This is the *binomial distribution*.

Example (continued)
From Table 1.3 above, the binomial distribution for the marginal probabilities of graduate studies for the whole class can be written

$$\Pr(n_{1\bullet} = 11, n_{2\bullet} = 17) = \binom{28}{11} \pi_1^{11} (1 - \pi_1)^{17}$$

whereas those for the two subgroups are, for males

$$\Pr(n_{11} = 5, n_{21} = 12 | \text{male}) = \binom{17}{5} \pi_{1|1}^{5} (1 - \pi_{1|1})^{12}$$

and for females

$$\Pr(n_{12} = 6, n_{22} = 5 | \text{female}) = \binom{11}{6} \pi_{1|2}^{6} (1 - \pi_{1|2})^{5}$$

Notice how the indices are used to distinguish among the various frequencies and probabilities.   □

In Section 1.3.2, we saw how to calculate the theoretical mean or expected value of a response variable. For fixed $n_\bullet$, n_1 is also a random quantity before we collect a sample. Thus, we can calculate its expected value using the probabilities given by Equation (1.9). The possible values of n_1 are $0, 1, \ldots, n_\bullet$; we can ignore the first of these because the probability in the sum will be multiplied by zero. Then,

$$
\begin{aligned}
E(n_1) &= \sum_{n_1} \Pr(n_1, n_2) n_1 \\
&= \sum_{n_1} \frac{n_\bullet!}{n_1!(n_\bullet - n_1)!} \pi_1^{n_1} (1 - \pi_1)^{n_\bullet - n_1} n_1 \\
&= \sum_{n_1} n_\bullet \frac{(n_\bullet - 1)!}{(n_1 - 1)!(n_\bullet - n_1)!} \pi_1^{n_1} (1 - \pi_1)^{n_\bullet - n_1} \\
&= n_\bullet \pi_1 \sum_{n_1} \frac{(n_\bullet - 1)!}{(n_1 - 1)!(n_\bullet - n_1)!} \pi_1^{n_1 - 1} (1 - \pi_1)^{n_\bullet - n_1} \\
&= n_\bullet \pi_1
\end{aligned}
$$

Notice that the sum in the second last line is for all possible binomial probabilities with a sample of size $n_\bullet - 1$ so that it equals one.

The theoretical variance of n_1 can also be calculated in a similar way and is equal to $n_\bullet \pi_1 (1 - \pi_1)$. The corresponding expressions for category i of a multinomial distribution are $n_\bullet \pi_i$ and $n_\bullet \pi_i (1 - \pi_i)$ for the mean and variance respectively.

Density and cumulative probability functions   As we saw in Section 1.2.3, if a variable is nominal, we can arbitrarily reorder its categories in creating a histogram. Such reordering is not possible for quantitative variables. In the latter case, the relative frequency polygon, which joins the $\hat{\pi}_i$, will usually have some fairly regular form.

One goal of statistical modelling is often to represent such a relationship theoretically by some simple function of the values of the quantitative variable, say
$f(y_i;\theta)$, where θ is some new unknown parameter replacing the set $\pi_1, \pi_2, \ldots, \pi_I$.
Thus, when this is reasonable, the multinomial distribution can be simplified by
using the relationship among the probabilities:

$$\pi_i = f(y_i;\theta)\Delta_i \qquad (1.10)$$

where Δ_i is the interval width or unit of measurement of y_i, as above in Section
1.2.3.

For a fixed value of θ, $f(y_i;\theta)$ gives the height of the probability curve at y_i,
in the same way as f_i in Equation (1.7) above. Thus, it is called the *probability
density function*. If an appropriate function can be found for the phenomenon
under study, the number of unknown parameters can then be reduced from $I -$
1, the π_i, usually to one or two, the θ. Such a result is especially important
because that functional form can often provide useful information about how the
data might have been generated. I shall discuss this in detail in Chapter 4.

Another closely related function which is often useful is the *cumulative distribution function*,

$$\begin{aligned}
\Pr(Y \leq y) &= \sum_{i \leq y} \pi_i \\
&= \sum_{i \leq y} f(i;\theta)\Delta_i \qquad (1.11) \\
&= F(y;\theta)
\end{aligned}$$

This gives the probability that the variable Y is less than or equal to some constant
value, y. For continuous variables, the sum is replaced by an integral.

1.4 Planning a study

In most scientific contexts, we are interested in studying some specific data generating mechanism. We assume some underlying phenomenon to operate to produce a response (distribution), perhaps different under different conditions. To
study this mechanism, we collect data. It is essential to carry this out in a way
that distorts that mechanism as little as possible, while still being feasible. Thus,
correctly planning or *designing* a study is extremely important. (See the quotation
from Fisher at the beginning of the Preface to the First Edition above.)

1.4.1 PROTOCOLS

Before beginning the study, a *protocol* should be developed to describe the steps
in obtaining and analysing the data. Among other things, it should clearly specify

(1) the question(s) to be investigated and how they relate to the theory under
    study;

(2) the population (or process) and time frame under consideration, and the unit of observation (person, family, village, ...);

(3) how the sample will be chosen, including the type of study design and the determination of sample size;

(4) what response and explanatory variables will be measured;

(5) what instruments will be used to measure the variables and how the people involved will be trained to use them;

(6) if certain variables will be under the experimental control of the research worker, how these will be assigned to individuals in the study;

(7) appropriate statistical models thought able to detect patterns of interest in the data to be collected and how they relate to the scientific theory under investigation;

(8) criteria to distinguish random from systematic variability;

(9) the form in which the final results will be reported.

When subsequently conducting the study, any violation of the protocol should only be allowed with strong justifications. All details of this must be recorded in a suitable amendment of the protocol.

1.4.2   OBSERVATIONAL SURVEYS AND EXPERIMENTS

Up until now, I have not always emphasised the difference between a sample and the population from which it comes. Indeed, in Chapter 2 when I begin to construct more realistic models, I shall ignore the distinction completely, but then concentrate on it in Chapter 3. I use this approach for didactic purposes of simplicity and clarity.

Nevertheless, in most situations, the distinction is essential in statistical modelling. We usually cannot observe all individuals in the population, and, thus, cannot calculate exactly the values of the parameters of a model. However, we do wish to obtain information about which model(s) would be most appropriate for the population, including some measure of what values the parameters of the model(s) might have, as well as an idea of the precision of such estimates. To do this, we look at a subgroup that is feasible to observe completely, called the sample, chosen from that population.

Sampling and inference   From the small group of individuals in a sample, we hope to be able to make *inferences* about how the models that we have constructed might work for the whole population. In other words, we shall wish to know how *likely* it is that models that work well for the sample will also be good for the whole population. This will involve several procedures. Among other things, we must

(1) calculate values of the unknown parameters for each reasonable model: *point estimation*;

(2) compare how well the various models explain the data: *model selection*; and

(3) for the better model(s), determine what range of values of the parameters is plausible: *interval estimation*.

We often retain several models and a range of parameter values for each, because, with only a subgroup of the whole population, we can never be sure about which is the best model or value. Thus, we shall try to design the study in such a way that

(1) appropriate information about the models of interest is available;

(2) the parameter estimates are as close as possible to the unknown values that would be obtained if the whole population could be observed, that is, are accurate; and

(3) the range of plausible values is as narrow as possible, that is, is precise.

However, before attempting to draw such inferences (Chapter 3), we shall need to know how to construct some useful models (Chapter 2).

Choosing a representative sample A major problem in the design of a sample is to choose a subgroup that is as representative as possible of the entire population. There is never one unique such subgroup. One fundamental way to accomplish this is by choosing a *simple random sample*: every member of the population, independently, has an equal probability of being selected for the sample. I have already discussed this possibility in Section 1.3.1. In order to do it, one must have a complete list of all members of the population from which to choose.

With a simple random sample, one has a good chance of a reasonable balance of the unknown characteristics, although this cannot be guaranteed. Most other more complex procedures are built upon this. Some improve the chance of representativity; others overcome the problem of requiring a complete list of population members. Here, I shall consider only two.

(1) If certain additional information is available about all the individuals in the population, increased representativity and precision may be obtained by *stratifying* the sample. This means choosing fixed proportions in each category of some known and readily available explanatory variable.

   Example
   If it is important that different age groups be properly represented, the whole population can be divided up by categories of age (if this information is available) and a simple random (sub)sample chosen separately from each age group. At least in the simplest case, the size of each such (sub)sample can be set proportional to the number of people of that age in the population.   □

   In this way, the sample and population proportions are guaranteed to be equal, at least for this stratifying variable.

(2) Time and expense may sometimes be saved, at the cost of reduced precision, by *clustering*, that is, choosing random groups of individuals found

together. The individuals are said to be nested in the groups. Such a procedure may also be useful when a complete list of clusters is available, but
not a complete list of members in the clusters.

Examples
In a cluster sample, several villages are chosen by simple random sampling
from a list of all villages in the population and then several people are selected from each chosen village, again by simple random sampling.

   Sometimes, the second stage of sampling is not performed. Entire classrooms of students may be chosen, where each classroom is taken at random
from the list of all classrooms.   □

If the variable used to specify the clusters is pertinent to the phenomenon
under study, individuals in each cluster will generally be more similar than
if they each had independently been chosen randomly from throughout the
whole population. Thus, less information will be available from a cluster
sample than from the same number of individuals chosen by simple random
sampling, but this should be compensated by increased feasibility.

An important distinction is between a *cross-sectional sample* design, which
makes observations at only one point in time, and a *longitudinal sample* design,
in which information is obtained from the same sample of individuals at several
points in time. I shall discuss this further in Section 1.4.3.

Causality and experimentation   Up until now, I have only considered taking a
sample from a population of interest and observing the existing values of certain
variables on the individual members of the sample. This may allow as to determine whether or not certain variables are associated with or dependent on each
other. However, often we wish to make stronger statements.

   When we observe the individuals in a sample at a given point in time, we may
often want to *predict* what we might observe if we chose another sample from
the same population. However, we must assume that we can do this under the
same circumstances; in other words, no change is taking place in the population
between the two sets of observations. This is static information. Even if we
observe the same sample over several time points, the information obtained only
refers to the evolution at those time points under those particular conditions. It
may be extrapolated to make a prediction into the future, but this will only be
useful if all the conditions of change remain fixed, as previously observed, and if
our theory (model) is appropriate.

   Neither of these procedures provides any information about what might occur
if a condition, an explanatory variable, is wilfully changed. If this occurs, the circumstances in which the sample was observed have been modified. Thus, when
we simply observe a sample, even over time, we do not know *why* each variable
changes. Although variables seem to change together, we gain no empirical in-

formation from the sample about which is influencing the other, or if either is. Of course, we may have theoretical information about the direction of this influence.

Example

In most normal circumstances, only an event that occurs strictly earlier in time can influence one later in time. In this way, we can exclude certain events, but this certainly does not imply that, if one event happens before another, the first necessarily influences the second! □

Suppose that our scientific theory tells us that two variables are *causally* related. In a deterministic situation, where no variability is present, causality implies that, by modifying the value of the first variable, we necessarily produce a specific subsequent change in the value of the second. This can often be assumed to occur in physics and chemistry, at least if we ignore measurement error.

However, when biological organisms are involved, variability will always be present. Then, one way to define causality operationally is to specify that, by modifying the value of the first variable, we only produce subsequent change in the *distribution* of the values of the second. In the presence of variability, causality can only, operationally, be applied to groups, and not to individuals. Notice that an explanatory variable such as sex (usually!) or age could not be a cause.

Example

Educational level is often thought to influence subsequent income. Suppose that we build a school in a village and that the children become better educated than if the school were not there. The subsequent distribution of the incomes of the young people, formerly students of the school, will be different than it would have been without the school. However, if we select one specific former student, his or her income might be the same as if the school had never existed. Of course, there is no way empirically to make such an observation about one individual. □

In contrast, to draw conclusions from observational samples, we must assume that the population remains the same or continues to evolve in the same way. Causality cannot be empirically studied simply by taking static samples from a population, even by following natural changes over time.

Example

Suppose that we administer a test to a sample of students. We record the scores obtained and the level of training of the teacher. We find that the mean score increases with higher levels of training of the teachers in the sample and draw the conclusion that the same kind of relationship holds in the population. This will be a valid description of the population as it exists.

However, it in no way means that increasing teacher training is the *cause* of higher mean scores. We cannot conclude that, if we improve the level of teacher training in the population, the students will have higher scores. It may well be that more highly trained teachers seek jobs in school with more intelligent children, that is, the higher mean scores at a school 'cause' the better trained teachers to be there. Or better equipped schools may be found in wealthier communities which attract, and can pay for, more highly trained teachers and which have children who do better in school. In other words, some external factors may be causing both higher mean score and better teacher training in the schools.   □

Instead of taking a sample of individuals from the population and observing the values of the variables that they have, suppose that we could select the individuals and then actually control them by giving them values of the variable(s) that we think are causes. This is called an *experiment*. The assignment of such values is usually done randomly, for the same reasons as in choosing a sample from a population, especially to eliminate biases.

Examples (continued)
To study the effect of education on income in the villages, we can choose a number of villages at random and construct new schools in half of them, chosen at random. Then, we subsequently compare the distributions of incomes of young people in the two subgroups of villages. In this way, we can observe how the distribution of incomes differs under the two conditions even though we cannot know what is the difference in effect for any specific individual.

To study the causal relationship between teacher training and students' results, we could choose teachers at random and give half, randomly, additional training. We then would assign children at random to their classes and record the scores on a test at the end of the year. We now would know the cause of the training: it is neither an external factor, such as community wealth, which also causes the mean scores, nor the scores themselves, but our own control.

However, in practice, such a study could only be carried out with the informed consent of both teachers and parents, something that would be unlikely to be given.   □

Refusals   In any study involving human beings, the people must agree to participate and must be aware of the implications. This is especially important in experiments where some changes will be imposed; the participants must be informed what the various treatments are and that they will be assigned randomly. Thus, in *any* study, people must always be able to refuse to participate. The attitude of statisticians to refusals has been somewhat different in observational surveys and in experiments. In the former, drastic measures are generally taken to reduce the refusals as much as possible whereas, in the latter, refusal after informed consent is accepted.

With every refusal, a random sample becomes less representative. The solution is not to replace the refusals by other similar individuals. A replacement individual who accepts to participate is necessarily different from the one who refused. Thus, because of the larger number of refusals in most experiments, the final sample of individuals involved, even if originally chosen randomly, will usually be less representative of the population of interest than in an observational survey. Drawing reasonable conclusions about the population of interest, called the *validity* of the study, will be more difficult.

Blinding   Ideally, in an experiment, especially with human beings, no person directly involved knows to what treatment group each individual belongs. This is necessary to avoid any unconscious differences in the ways in which the groups are handled or influences on the way they react. It is known as a *blinded design*.

Example
In the teacher training example, the teachers necessarily know whether or not they receive extra training. On the other hand, the parents and children should not know whether or not their teacher had the training. Otherwise, this could influence their reactions. Those knowing that their teachers did not receive extra training might work extra hard in order to compensate, in this way reducing the difference between the two groups.

In contrast, suppose that two types of drugs are to be compared in the treatment of some disease. Then, neither patients nor doctors should know who receives which drug. If a drug is compared to no treatment, those not receiving the drug are given an inert compound called a *placebo* that looks and tastes identical to the real thing. This is a *double blinded design*.   □

In studies involving human beings, causality is, thus, very difficult to ascertain empirically, which is not to belittle its extreme importance.

Example
Think of the relationship between smoking and lung cancer. The debate lasted for many years, although sampling from existing populations showed an association whereby proportionally more smokers had lung cancer. But an experiment could not be performed in which some people were (randomly chosen and) told to smoke and others not, after which cancer incidence would be observed in the following years.   □

Thus, both in a static sample and in an experiment, we may find a relationship of dependence between two variables. The statistical model used to represent the relationship may be the same in both cases. But the conclusions about the meaning of the relationship will depend on the way in which the information was collected. No mathematical manipulation of the data afterwards can change this. Causality, as defined above, can only be studied empirically if the appropriate explanatory variables can be manipulated. Causal conclusions can only be drawn from an observational sample, without experimentation, by making empirically unverifiable assumptions.

Such problems are specific to studies of human beings. In most other situations, experiments can be performed instead simply of observational studies, although this is not always possible, for example in ecology or astronomy. Animals or plants involved in biological studies are not generally given the possibility of refusing. Material in chemical and physical experiments may be uniform enough that random sampling from a population is not necessary, although random allocation to the treatments is always a good idea.

1.4.3   STUDY DESIGNS

The data to answer most scientific questions cannot simply be obtained by rushing out and collecting information from some simple random sample. Considerable care must be taken in planning how the complete study is to be organised and conducted. This should be clearly set out in the protocol (Section 1.4.1) and then closely followed. Let us now look at possible *designs* for such studies.

Prospective designs   In a *prospective design*, individuals are sampled from a population and then followed over a certain period of time. At least three important cases may be distinguished:

(1) In a *cohort* or *panel design*, all variables are simply observed as they occur over time.
(2) In a *clinical trial*, often used in medical studies, the available and consenting subjects, not randomly chosen, are randomly assigned to one of a number of different treatments before the follow-up observations.
(3) In a *capture-recapture design*, a sample is obtained and marked in some distinctive way that hopefully does not influence the phenomenon under study. At some subsequent time point, a second sample is taken and information gained from the fact that some in it have the mark from the first sampling.

Example
A study was made of migration among four areas of Britain, the Metropolitan Counties, between 1966 and 1971, as given in Table 1.8. (Moves to and from other areas are not included in the table.) Although this was based on the published migration reports of the 1971 British census, the information comes from a panel study. It used census information on a 10% sample on those dates, with the same variable being observed twice. On the left, we see the place of residence of each individual in 1966, whereas on the top we see the same list for residence in 1971.

We can immediately notice the large diagonal frequencies, indicating that many people had their residence in the same area in both years. However, this does not necessarily mean that they did not change residence in that period. Some could have moved house within an area or moved out of an area and back in within the period.   □

Of the designs described here, the clinical trial is the only one that is experimental. However, many such trials, especially those conducted by private firms,

Table 1.8. Geographical migration among areas of Britain between 1966 and 1971. (Fingleton, 1984, p. 142)

| | 1971 | | | | |
1966	Central Clydesdale	Lancashire & Yorkshire	West Midlands	Greater London	Total
Central Clydesdale	118	12	7	23	160
Lancashire & Yorkshire	14	2127	86	130	2357
West Midlands	8	69	2548	107	2732
Greater London	12	110	88	7712	7922
Total	152	2318	2729	7972	13171

are not used for scientific purposes but rather as an aid in making a decision as to whether or not some new medical treatment can be marketed.

Among clinical trials, two distinct procedures may be used:

(1) One treatment is assigned at random to each participant and they are all followed over time until some endpoint specified in the protocol. This is called a *parallel group* trial.

(2) Two or more different treatments are randomly assigned to each participant, in different orders. Each treatment is applied for a specified length of time and the result recorded. Then, perhaps after a washout period so that the effect of the previous treatment disappears, the individual switches to the next treatment, and so on. This is a *cross-over* trial.

The advantage of a cross-over design is that the various treatments are compared on the same individuals so that any differences in treatment effect are not confounded with individual differences. However, many types of treatments are not amenable to being switched in this way.

Cross-sectional designs   A *cross-sectional design* simply observes all variables on individuals at one given fixed point in time.

Example
Injuries in car accidents in Florida, USA, in 1988 were compiled by the Department of Highway Safety and Motor Vehicles in that state. They were classified, in Table 1.9, as to whether a seatbelt was being used at the time or not.   □

If a study is carried out over time, but different individuals are involved at each time point, this is a series of cross-sectional studies, not a prospective study.

Table 1.9. Car accidents in Florida, USA, in 1988, classified by whether or not a seatbelt was worn. (Agresti, 1990, p. 30)

	Injury		
Seatbelt	Fatal	Nonfatal	Total
No	1601	162527	164128
Yes	510	412368	412878
Total	2111	574895	577006

Retrospective designs   In a *retrospective* design, subjects are selected and their present values of any relevant variables are recorded. Then, the values of any other necessary explanatory variables are obtained by examining the previous history of the individuals. Most so-called cross-sectional studies are actually retrospective because many of their explanatory variables come from the past history of the individuals involved.

Example
Working people in a random sample are interviewed to obtain their opinion on the present government. This is to be related to their education level, a variable that was determined some time previous to the study.   □

A special case of a retrospective design is the *case-control* design in which subjects are actually chosen according to their values of the response variable. Then, pertinent explanatory variables, called *risk factors*, are sought in their previous history. Thus, in this design, the explanatory variables are random and the response fixed. This approach is often used when one of the response events is uncommon, as for a rare disease, because a prospective study would require an enormous sample in order to obtain even a few individuals with that event. In such a design, the proportions of the possible response events in the sample are often equal, very different from those in the population of interest. In this sense, the sample is not at all representative. Very special models are thus necessary (Sections 2.1.4 and 2.3.1).

Example
A study was made to investigate the effect of oral contraception use on the chance of a heart attack. 58 married women less than 45 years old and under treatment for myocardial infarction in two hospital regions of England and Wales during 1968–1972 were each *matched* with three control patients in the same hospitals who were being treated for something else, as shown in Table 1.10. (Agresti does not state why there are 8 too few control patients.) Matching means that patients were chosen who had known characteristics as similar as possible to those under treatment. Subjects were, then, asked if they had ever previously used contraceptives. Obviously, one in four women, the proportion in this study, would not be expected to have a myocardial infarction in the whole population of interest.   □

Table 1.10. Retrospective study of myocardial infarction as depending on contraceptive use. (Agresti, 1990, p. 12, from Mann *et al.*)

	Myocardial infarction		
Contraceptive	Yes	No	Total
Yes	23	34	57
No	35	132	167
Total	58	166	224

Both retrospective and prospective studies are *longitudinal designs*. They can provide information about processes over time. However, as we have seen, not all types of designs allow us to draw the same types of conclusions, in particular those about the causal effect of one variable on another.

One further important aspect of design, that I have not yet discussed, is the decision as to how many individuals are to be involved, called the *sample size*. This will depend on the models to be fitted so that I must leave description of how it can be calculated to the following chapters.

1.5 Exercises

(1) Give two examples of each of the types of variables described in Section 1.1.2, nominal, ordinal, integral, and continuous.
  (a) How many possible different values does each have?
  (b) For each variable, give the unit of measurement.
  (c) Which may present problems in obtaining accurate results?
  (d) Which do you think can be observed most precisely?
  (e) For each, what will be the most appropriate way of summarising some observed data?

(2) What are the standard errors of the empirical variance and of the empirical standard deviation of a set of $n_\bullet$ observations?

(3) (a) Show that the theoretical variance of n_1 in the binomial distribution is equal to $n_\bullet \pi_1(1 - \pi_1)$.
  (b) Derive the theoretical mean $n_\bullet \pi_i$ and variance $n_\bullet \pi_i(1 - \pi_i)$ of n_i for the multinomial distribution.

(4) In Table 1.7, calculate
  (a) the means and
  (b) the standard deviations
  before and after diet.

(5) (a) Calculate appropriate cross-classified percentages for the following data:
    i. the migration data of Table 1.8;
    ii. the car accident data of Table 1.9;
    iii. the myocardial infarction data of Table 1.10.

(b) Give the estimated joint, marginal, and conditional probabilities for these tables.

In each case, discuss any relationships that may be apparent.

(6) Data were collected in a study of the relationship between life stresses and illnesses. One randomly chosen member of each randomly chosen household in a sample from Oakland, California, USA, was interviewed. In a list of 41 events, respondents were asked to note which had occurred within the last 18 months. The results given are for those recalling only one such stressful event. The classification variable is the number of months prior to an interview that subjects remember a stressful event. Thus, the following table gives the frequency of recall of one stressful event in each of the 18 months preceding an interview (Haberman, 1978, p. 3).

Month	1	2	3	4	5	6	7	8	9
Respondents	15	11	14	17	5	11	10	4	8
Month	10	11	12	13	14	15	16	17	18
Respondents	10	7	9	11	3	6	1	1	4

Make a percentage table and a histogram of these results.

(7) The following two tables give the observed frequencies of some (unfortunately) unspecified type of accidents (Skellam, 1948)

Accidents	Frequency
0	447
1	132
2	42
3	21
4	3
5	2

and of car accidents in a year for 9461 Belgian drivers (Gelfand and Dalal, 1990, from Thyrion).

Accidents	Frequency
0	7840
1	1317
2	239
3	42
4	14
5	4
6	4
7	1

(a) Calculate the percentage tables for the two sets of frequencies.

(b) Plot the histograms and compare them.

(c) Discuss whether the first table might also refer to car accidents, keeping in mind the lapse of time between the publication of the two sets of data.

(8) The table below shows the numbers of units of two types of consumer goods purchased by 2000 households over 26 weeks (Chatfield *et al.*; the frequency for 21 units in the last column refers to > 20). The two studies were separated in time by about seven years.

Units bought	Number of households buying	
	Item A	Item B
0	1612	1498
1	164	81
2	71	47
3	47	25
4	28	16
5	17	17
6	12	6
7	12	10
8	5	3
9	7	3
10	6	6
11	3	4
12	3	4
13	5	3
14	0	2
15	0	2
16	0	3
17	2	1
18	0	0
19	0	2
20	1	1
21	0	12
22	2	
23	0	
24	0	
25	1	
26	2	

(a) Why might there seem to be a somewhat larger number of people buying about 13 or 26 items?

(b) Calculate the percentage tables for the two sets of frequencies.

(c) Plot the histograms and compare them.

(9) (a) Plot the data in Exercise (1.6) above as points on a scattergram.

(b) Does this suggest any other interpretation for these data than that from the histogram produced in that exercise?

(10) Consider the following two study designs.

(a) A simple random sample is drawn from women visiting a birth control clinic for the first time. They are asked whether or not they use contraceptives.

(b) A simple random sample is drawn from the list of all divorces granted in a large city over a year's time. For each couple, the length of marriage is recorded.

In each case,

(a) Describe carefully for what larger population inferences may be drawn.

(b) Give the major drawbacks of each design.

(c) Explain how you would improve the design.

2
Categorical data

Observations that involve only nominal or ordinal variables (Section 1.1.2) are called categorical data. In this chapter, I shall concentrate on the special methods of model construction for such data, and specifically on the case when the *response* variable is categorical. These models have the major advantage that few assumptions are made in constructing them. On the other hand, when possible, stronger and more informative conclusions can usually be drawn from models constructed using more assumptions, if they are appropriate. However, I shall delay presentation of models incorporating such further assumptions until Chapter 4.

As we saw in Section 1.2.1, often categorical data can be summarised in contingency tables. Only if all the variables used to create such a table are categorical, and the original categories from the raw data are used, will this involve no loss of information about the relationships among those variables. Thus, in this chapter, we shall work with a condensation of the raw data in the form of contingency tables.

However, we also saw that *all* empirically observed data must fall into discrete categories, even if the variable is theoretically continuous. Thus, the methods for categorical data to be presented here will, in fact, also be applicable to any type of variable. The values of all observable variables are limited in precision; they have categories determined by their units of measurement. However, as we shall see in later chapters, this will not always necessarily be the best approach for quantitative, that is, integral and continuous, variables.

2.1 Measures of dependence

In Section 1.3.2, we saw how to detect independence between two variables. Either of the two equivalent relationships, Equation (1.5):

$$\Pr(y_1|x_1) \overset{?}{=} \Pr(y_1)$$

or Equation (1.6):

$$\Pr(y_1 \text{ and } x_1) \overset{?}{=} \Pr(y_1)\Pr(x_1)$$

can be examined; the equalities only hold if the two variables are independent. Intuitively, it is easier to understand the first: if the response variable is independent of the explanatory variable, the conditional probability of the response, given the value of the explanatory variable, must equal the marginal probability. Knowing the value of the explanatory variable does not help to predict what will be the value of the response.

Unfortunately, when independence does not hold, these relationships do not help us in measuring the *strength* of dependence. For simplicity, let us continue to look at relationships between only two variables for the moment.

2.1.1 ESTIMATION

Before proceeding, let us recall how to calculate or *estimate* probabilities when we have a sample of data available.

In Section 1.3.1, we saw that a simple probability, related to one (response) variable, in a population is defined by

$$\pi_i = \frac{n_i}{N}$$

where n_i is the number of individuals in the population with category i and N is the total size of the population. Generally, we cannot actually calculate such values because we cannot observe the whole population.

However, we also saw that, if we only observe a sample from the population, an *estimate* of that probability can be obtained from the corresponding relative frequencies:

$$\widehat{\pi_i} = \frac{n_i}{n_\bullet}$$

where n_i and $n_\bullet$ now refer to the sample, and not to the population. Now, we need to extend these ideas to probabilities with two (or more) variables.

Notation   In Section 1.2.1 and Table 1.4, I introduced abstract notation to describe the frequencies in a two-way contingency table, based around n_{ij} and various sums. Similar notation will be useful for manipulating probabilities.

Consider any two variables, Y and X, with observable values indexed, respectively, by i and j. Let us denote by π_{ij} the joint probability of any pair of outcomes (y_i, x_j), that is, the categories (i, j) occurring. The set of such probabilities describes the joint distribution of Y and X. Their sum must equal one:

$$\sum_i \sum_j \pi_{ij} = 1$$

Then, the marginal distributions are obtained by summing the joint probabilities across rows or columns. In a similar way to what I did with the frequencies, I shall denote these by

$$\pi_{i\bullet} = \sum_j \pi_{ij}$$

$$\pi_{\bullet j} = \sum_i \pi_{ij}$$

The second relationship above for independence, Equation (1.6), can now be written

$$\pi_{ij} = \pi_{i\bullet}\pi_{\bullet j}$$

The marginal probabilities only allow us to calculate the joint probability if there is no relationship between the variables. The marginal probabilities contain no information about the dependence between two variables. Only the joint and the conditional probabilities contain such information.

The joint distribution is symmetric in the two variables. Thus, it is primarily useful when we do not make a distinction between response and explanatory variables, but simply wish to examine the reciprocal dependence between two variables. I shall discuss this further in Section 2.3.2.

Usually, we want to understand how one special variable, the response, depends on one or more other variables, the explanatory variables. (Recall that I shall generally take i to index the response variable and $j, k, \ldots$ the explanatory variables.) When one variable Y is taken to be a response and the other, X, an explanatory variable, then we no longer have a symmetric situation. Now, we assume that Y is random, but X is fixed. The joint distribution is no longer meaningful. Instead, we should use the conditional distribution of Y for fixed and known X, with probabilities

$$\pi_{i|j} = \frac{\pi_{ij}}{\pi_{\bullet j}}$$

When independence does not hold, we shall wish to compare the conditional distribution of Y for various values of the explanatory variable X.

The estimates from the data in a sample will be

$$\widehat{\pi_{ij}} = \frac{n_{ij}}{n_{\bullet\bullet}}$$

for the joint distribution,

$$\widehat{\pi_{i\bullet}} = \frac{n_{i\bullet}}{n_{\bullet\bullet}}$$

$$\widehat{\pi_{\bullet j}} = \frac{n_{\bullet j}}{n_{\bullet\bullet}}$$

for the marginal distributions, and

$$\widehat{\pi_{i|j}} = \frac{\widehat{\pi_{ij}}}{\widehat{\pi_{\bullet j}}}$$

$$= \frac{n_{ij}}{n_{\bullet j}}$$

for the conditional distribution.

Table 2.1. The fixed margins of a contingency table.

X	y_1	y_2	Total
	\multicolumn{2}{c}{Y}	
x_1			$n_{\bullet 1}$
x_2			$n_{\bullet 2}$
Total	$n_{1 \bullet}$	$n_{2 \bullet}$	$n_{\bullet \bullet}$

Table 2.2. The fixed margins of a contingency table, from Table 1.3.

Sex	Yes	No	Total
	\multicolumn{2}{c}{PhD plans}	
Male	3		17
Female			11
Total	11	17	28

Example (continued)
For the graduate studies plans in Table 1.3, take PhD plans as the response variable and sex as the explanatory variable. The estimates are (0.18, 0.43, 0.21, 0.18) for the joint probabilities, (0.61, 0.39) for the marginal probabilities for sex, (0.39, 0.61) for the marginal probabilities for plans, (0.29, 0.71) for the conditional probabilities of plans given male, and (0.55, 0.45) for the conditional probabilities given female. The latter two sets were already given in Table 1.5 and Figure 1.4, the former as percentages.   □

Fixing margins   In a contingency table, the simple marginal totals describe the overall structure of the population. In constructing models, they are usually taken to be fixed, in the same way that an explanatory variable is assumed fixed in a conditional probability. This assumption is usually made even though they were not fixed in the original design of the study. A design fixes a margin if it stratifies on the variable for that margin (Section 1.4.2); in a case-control study, the margin for the response is fixed (Section 1.4.3). Thus, in the study of the relationship among variables, we assume the structure defined by the margins to remain constant; we condition on the observed margins.

Suppose now that we have a blank table, but with given marginal totals, as shown in Table 2.1. Given that the frequencies inside the table must sum to the fixed marginal totals, only a certain number of them can be arbitrarily entered into such a blank table before the table is completely determined.

Example (continued)
The margins from Table 1.3 are shown here in Table 2.2. Let us arbitrarily set $n_{11} = 3$. Notice that this value must lie between 0 and 11 if the marginal con-

straints are not to be violated. Once this value is placed in the table, all the other values are determined and no more can be supplied arbitrarily.   □

Thus, in a 2×2 table, only one such arbitrary value can be entered. Then, all the others are automatically given. The number of frequencies that can be arbitrarily entered before the complete table is fixed is called the number of *degrees of freedom* (d.f.) of the table. Thus, for a 2×2 table, there is one degree of freedom. A little thought should show that, for a two-way table of size $I \times J$, the number of degrees of freedom is given by $(I - 1) \times (J - 1)$.

2.1.2 INDEPENDENCE

Let us now rewrite the two definitions of independence in this notation. As we just saw above, from the product law of Section 1.3.2, the variables X and Y are statistically independent if all joint probabilities equal the product of the corresponding marginal probabilities:

$$\pi_{ij} = \pi_{i\bullet}\pi_{\bullet j} \qquad \text{for all } i, j \tag{2.1}$$

This is Equation (1.6). As we also saw in Section 1.3.2, this is equivalent to the conditional relationship,

$$\pi_{i|j} = \pi_{i\bullet} \qquad \text{for all } i, j \tag{2.2}$$

This is Equation (1.5). Each conditional distribution of Y, for any fixed $X = x_j$, is equal to the marginal distribution. Thus, under independence, when the probabilities are the same for all possible fixed conditions X, the response Y does not depend on those conditions.

Examples (continued)
For graduate plans for doing a PhD, in Table 1.3, we saw, both above and in Section 1.3.2, that the conditional probabilities are 0.29 for male and 0.55 for female students, indicating a dependence on sex.

In the car accident example of Table 1.9, the type of injury might be taken as a response, given the fact that a seatbelt was being worn at the time or not. The conditional probability of a fatal accident, given that a seatbelt was worn, is estimated as $\widehat{\pi_{1|2}} = 0.0012$ compared with $\widehat{\pi_{1|1}} = 0.0098$ without a seatbelt. From Equation (2.2), the difference indicates a dependence of type of accident on whether a seatbelt was worn or not.

In a social survey in the USA, called the 1982 General Social Survey, people were asked about their opinions on the death penalty and on gun registration, with the results in Table 2.3. The estimates of the joint probabilities are $(0.56, 0.17, 0.22, 0.05)$ and of the marginal probabilities, $(0.73, 0.27)$ for gun registration and $(0.78, 0.22)$ for the death penalty. There appears to be no reason to assume that either variable should be taken as a response with the other fixed, so that I shall look at the joint probabilities. Under independence, they are estimated as $(0.57,$

Table 2.3. Opinions on gun registration and the death penalty. (Agresti, 1990, p. 29, from Clogg and Shockey)

Gun Registration	Death penalty		Total
	Favour	Oppose	
Favour	784	236	1020
Oppose	311	66	377
Total	1095	302	1397

Table 2.4. A deterministic contingency table.

Sex	PhD plans		Total
	Yes	No	
Male	0	17	17
Female	11	0	11
Total	11	17	28

0.16, 0.21, 0.06), as compared with those just given above. Because these are not identical, from Equation (2.1) at least some small dependence is indicated.   □

Only if there is dependence does the distribution of the response variable change among the various categories of the explanatory variable(s). One may, however, wonder just how much change is possible. What extreme cases are possible?

Examples
Let us consider two imaginary cases based on the PhD plans in Table 1.3, each having the same marginal totals. Above, we saw that we can only choose one arbitrary entry in a 2 × 2 table when the margins are fixed.

The first extreme case is shown in Table 2.4. Although this table has the same marginal totals as the original table, here if we know the sex of an individual, we know the graduate plans. This is a deterministic table, with probabilities of zero and one. Here, we have the largest possible change in conditional probability between the two sexes.

Next, let us create another table, again with these same marginal totals, but this time, applying the independence formula of Equation (2.1). We shall use it to predict what values would be found in the table under this assumption. These are called the *fitted values* under the assumption of independence. Thus, we have

$$\Pr(x_1)\Pr(y_1) = \widehat{\pi_{1\bullet}}\widehat{\pi_{\bullet 1}}$$
$$= \frac{11}{28} \times \frac{17}{28}$$
$$= 0.24$$

This is the proportion of all 28 individuals in the table who should be in the upper

Table 2.5. A contingency table with independence.

	PhD plans		
Sex	Yes	No	Total
Male	6.68	10.32	17
Female	4.32	6.68	11
Total	11	17	28

left cell of the table if independence holds. This means that we should now have $n_{\bullet\bullet}\widehat{\pi}_{11} = n_{\bullet\bullet}\widehat{\pi}_{1\bullet}\widehat{\pi}_{\bullet1} = 28 \times 0.24 = 6.68$ individuals in that cell.

As we know, fixing this value determines the contents of the entire table, which is given in Table 2.5. Obviously, this can only be imaginary, because frequencies must be integers. Nevertheless, let us calculate the conditional probabilities, $\widehat{\pi}_{1|1} = 6.68/17 = 0.39$ and $\widehat{\pi}_{1|2} = 4.32/11 = 0.39$. Because we have constructed a table with independence, the probabilities of planning to go on to do a PhD are now identical for both sexes. This is another extreme, where both conditional probabilities are identical.   □

2.1.3 COMPARISON OF PROBABILITIES

As we saw in Section 1.3.2 and discussed again above, Equations (1.5) and (1.6), or (2.1) and (2.2), for independence provide no useful way for us to measure the strength of dependence when variables are not independent. The question is how we can usefully compare different probabilities. Let us examine several possibilities. The intuitive ways are not always the best!

Differences   For the conditional probabilities, any two rows, say j and l, of a table can be compared by taking the appropriate differences of probabilities: $\pi_{i|j} - \pi_{i|l}$. Such differences must lie between -1.0 and 1.0; we just saw one example of such an extreme value (in Table 2.4). If all differences are zero, the conditional probability distributions are identical and the two variables are independent. The farther the differences are from zero, the greater the dependence.

The drawback of this rather intuitive approach is that a difference in probabilities of given size may have greater importance when the proportions are close to the limits, 0 or 1, than in the middle, near 0.5.

Example
Suppose that one of the conditional probabilities is 0.5. Then, the greatest difference that a smaller conditional probability can have is 0.5. However, if the conditional probability is 0.1, this difference is now at most 0.1. Thus, the differences are influenced by how close the probabilities are to 0 or 1. Differences can be greater if one of the probabilities is near 0.5 than if both are near the extremes. One may ask whether or not $0.5 - 0.4$ can be interpreted in the same way as $0.15 - 0.05$.   □

Although statistical models based on this approach were used in the past, in most situations, they are no longer considered to be acceptable.

Ratios and relative risk   A second approach is to use the ratio of two conditional probabilities under different conditions, $\pi_{i|j}/\pi_{i|l}$. This is known as the *relative risk*. It can take any non-negative value. If all relative risks are equal to unity, the variables are independent.

Example (continued)
In the car accident example of Table 1.9, the relative risk of a fatal accident is estimated to be

$$\frac{\widehat{\pi_{1|1}}}{\widehat{\pi_{1|2}}} = \frac{1601/164128}{510/412878}$$
$$= 7.90$$

when not wearing a seatbelt as compared with wearing one. Nonseatbelt wearers have a much higher risk of fatal accident than seatbelt wearers. This indicates a dependence of the outcome of the accident on whether or not a seatbelt was worn. Because the table does not include injury-free accidents, this statement is also conditional on there being an injury.   □

In many situations, a disadvantage of relative risk comparisons is that they use only one category of the response variable.

Example (continued)
For a nonfatal accident, the relative risk is

$$\frac{\widehat{\pi_{2|1}}}{\widehat{\pi_{2|2}}} = \frac{162527/164128}{412368/412878}$$
$$= 0.99$$

In contrast to the result above, here there is little difference in relative risk for nonfatal accidents. This seems to indicate a lack of dependence of the outcome on whether or not a seatbelt was worn, contradicting that above. However, when examined closely, both values make sense.   □

Thus, this measure of dependence differs depending on which category of the response variable is used. Although it provides useful information, it does not provide an adequate measure of dependence.

Relative risk is widely used in medical applications where only one category of response, usually related to a specific disease, is of interest. The relative risk of say dying from a disease under different conditions, such as two competing treatments, is calculated. Only the relative risk of dying from the disease, and not that of surviving, under those conditions is of interest.

Odds ratio The relative risk involves the ratio of two conditional probabilities of the same response value under two different conditions defined by values of the explanatory variable. A second possible ratio is that of two probabilities of different response values under the same conditions. This is known as the *odds*. For the moment, let us concentrate on the case when the response has only two possible values. Then, the odds are

$$\frac{\pi_{1|j}}{\pi_{2|j}} = \frac{\pi_{1|j}}{1 - \pi_{1|j}}$$
$$= \frac{\pi_{1j}}{\pi_{2j}}$$

Notice how they can be defined, equivalently, either by the conditional or by the joint probabilities. The odds of one event versus another happening are, thus, defined as the ratio of probabilities of the two events. Such a ratio can take any non-negative value. If the odds are greater than unity, then response category 1 is more probable than category 2, under conditions j, and conversely. For independence, the set of odds under each condition, j, must be the same.

In horse racing, the odds of a given horse losing are described as, say, three to one. This means that the ratio of the probability of losing to winning is $3/1$, so that the probability of winning is $1/4$.

A special characteristic of the odds is that their estimation will not involve the marginal totals. This is because the estimates of the two probabilities in the ratio will necessary have the same denominator:

$$\frac{\widehat{\pi_{1|j}}}{\widehat{\pi_{2|j}}} = \frac{n_{1j}/n_{\bullet j}}{n_{2j}/n_{\bullet j}}$$
$$= \frac{n_{1j}}{n_{2j}}$$

Example (continued)
In the car accident example of Table 1.9, the odds of a fatal as compared with a nonfatal injury is estimated to be

$$\frac{\widehat{\pi_{1|1}}}{\widehat{\pi_{2|1}}} = \frac{n_{11}}{n_{21}}$$
$$= \frac{1601}{162\,527}$$
$$= 0.0099$$

without a seatbelt and

$$\frac{\widehat{\pi_{1|2}}}{\widehat{\pi_{2|2}}} = \frac{n_{12}}{n_{22}}$$

$$= \frac{510}{412\,368}$$

$$= 0.0012$$

with one. Again, this indicates a dependence of type of accident on whether or not a seatbelt was worn. □

The log odds

$$\log\left(\frac{\pi_{1|j}}{\pi_{2|j}}\right) = \log\left(\frac{\pi_{1|j}}{1 - \pi_{1|j}}\right)$$

is often called the *logit*. As usual in statistics, logarithms are Naperian or natural logs. These are most often denoted by 'ln' on hand calculators but I shall use the standard notation, 'log', here.

The ratio of two odds is called the *odds ratio* or *cross product ratio*. Consider the case when the explanatory variable also has only two categories:

$$\frac{\pi_{1|1}}{\pi_{2|1}} \bigg/ \frac{\pi_{1|2}}{\pi_{2|2}} = \frac{\pi_{1|1}\pi_{2|2}}{\pi_{2|1}\pi_{1|2}}$$

$$= \frac{\pi_{11}}{\pi_{21}} \bigg/ \frac{\pi_{12}}{\pi_{22}}$$

$$= \frac{\pi_{11}\pi_{22}}{\pi_{21}\pi_{12}}$$

This again can take any non-negative value. Degrees of dependence are measured from unity, which indicates independence. A value greater than unity indicates the same degree of dependence, but in the opposite direction, as its reciprocal, which will be less than unity. Thus, the ranges are not symmetric, being $(1, \infty)$ above one and $(0, 1)$ below. However, the odds ratio is symmetric in the variables, as can be seen from its definition in terms of joint probabilities.

This symmetry in the two variables is an important characteristic of the odds ratio. This can be seen most clearly when it is expressed in terms of the ratio of joint probabilities. Thus, if, for some reason, we conditioned the probabilities for the explanatory variable on the values of the response, we would obtain the same value as the usual conditioning of response on explanatory variable.

Because estimation of the odds does not involve the marginal totals, the same is true of the odds ratio:

$$\frac{\widehat{\pi_{1|1}}\widehat{\pi_{2|2}}}{\widehat{\pi_{2|1}}\widehat{\pi_{1|2}}} = \frac{n_{11}n_{22}}{n_{21}n_{12}}$$

Example (continued)
In the car accident example of Table 1.9, the estimated odds ratio is

$$\frac{n_{11}/n_{21}}{n_{12}/n_{22}} = \frac{1601/162527}{510/412368}$$
$$= 7.96$$

Interchanging variables, we still obtain

$$\frac{n_{11}/n_{12}}{n_{21}/n_{22}} = \frac{1601/510}{162527/412368}$$
$$= 7.96$$

□

As for the logit instead of odds, often it is more convenient to use the *log odds ratio*,

$$\log\left(\frac{\pi_{1|1}\pi_{2|2}}{\pi_{2|1}\pi_{1|2}}\right) = \log\left(\frac{\pi_{11}\pi_{22}}{\pi_{21}\pi_{12}}\right)$$

This can take any value and is symmetric in measuring dependence on each side of independence, which is at 0.

Example (continued)
In the car accident example of Table 1.9, the estimated odds ratio is 7.96 and the corresponding log odds ratio, 2.075. Both indicate a positive dependence between fatal injuries and not wearing a seatbelt, that is, that there is a much greater chance of a fatal accident without a seatbelt.   □

However, one major problem with any ratio of probabilities, such as relative risk or odds, is that its estimate is infinite if a probability in the denominator is estimated to be zero. The log odds has a similar problem if any probability is estimated to be zero, as in Table 2.4 above.

2.1.4   CHARACTERISTICS OF THE ODDS RATIO

Odds ratios have a number of characteristics that make them attractive as a basis for constructing statistical models for categorical response data.

Retrospective designs   As we have seen, the odds ratio is symmetric in the variables and its estimation does not involve the marginal totals. As a result of these characteristics, it has a further useful property. It can measure dependence even when the study is performed 'backward', as in a retrospective or case-control study (Section 1.4.3). There, the marginal distribution of the response variable is fixed by the design.

Example (continued)
The myocardial infarction data of Table 1.10 arise from a case-control design. Thus, the marginal distribution of the response variable, myocardial infarction, was fixed by this design. However, we can still estimate the odds ratio; it is

$$\frac{23 \times 132}{34 \times 35} = 2.55$$

The dependence of infarction on contraceptive use, as measured by the log odds ratio, is

$$\log\left(\frac{23 \times 132}{34 \times 35}\right) = 0.937$$

indicating a strong positive relationship between them.   □

Relation to relative risk   For a 2×2 table, we have

$$\frac{\pi_{1|1}}{\pi_{1|2}} \times \frac{\pi_{2|2}}{\pi_{2|1}} = \frac{\pi_{1|1}}{\pi_{1|2}} \times \frac{1 - \pi_{1|2}}{1 - \pi_{1|1}}$$

The first factor is the relative risk for the first category of response. If the conditional probability of response one, $\pi_{1|j}$, is small for both groups, the second factor will be close to unity and the relative risk and odds ratio will be very similar.

Example (continued)
In the car accident example of Table 1.9, the conditional probabilities of fatal injury are 0.0099 for nonseatbelt wearers and 0.0012 for seatbelt wearers, both very small. Recall that we found the odds ratio to be 7.96 and the relative risk of a fatal accident to be 7.90, very similar values.   □

This result is especially important in retrospective designs where the appropriate conditional probability estimates are not available, so that the relative risk cannot be directly estimated.

Example (continued)
In the data of Table 1.10, we cannot calculate the relative risk because the proportions of cases of myocardial infarction and controls are fixed by the design. However, we just saw that the odds ratio is estimated to be 2.55. If myocardial infarction is rare in the population of women under study, this will be an approximate estimate of the relative risk of infarction for contraceptive users as compared with nonusers.   □

$I \times J$ **tables**   In a 2×2 table, all four possible odds ratios are simply permutations of the frequencies in the numerator and denominator. For larger tables, a

Table 2.6. Global results for testing a new medical treatment.

Treatment	Recover Yes	No	Total	Recovery rate
New	20	20	40	50%
Old	16	24	40	40%

number of distinct odds ratios can be calculated. In a $I \times J$ table, there will be $(I-1)(J-1)$ *local* odds ratios obtained by comparing adjacent categories of the two variables:

$$\frac{\pi_{ij}\pi_{i+1,j+1}}{\pi_{i,j+1}\pi_{i+1,j}}, \qquad i=1,\ldots,I-1, j=1,\ldots,J-1$$

These determine all possible odds ratios and contain all the information in them.

However, the construction of a minimal set of odds ratios is not unique. Another possibility would be to make comparisons with respect to some special category, for example, the first:

$$\frac{\pi_{11}\pi_{ij}}{\pi_{1j}\pi_{i1}}, \qquad i=2,\ldots,I, j=2,\ldots,J$$

Notice that again there are $(I-1)(J-1)$ distinct ratios. The set that can most usefully be interpreted for a given problem should be used.

2.1.5 SIMPSON'S PARADOX

We have now looked at a number of ways in which the dependence between two variables can be measured. However, care must be taken as to the interpretation of such measures. It is often possible to find relationships among variables that make no sense, such as that between tobacco production in Cuba and divorces in the USA. As we have seen in Section 1.4.2, the exact interpretation of dependence will depend on the way in which the data were originally collected. Only in certain special circumstances can one draw conclusions in terms of causality between the variables. As an illustration of the danger, consider the following example.

Example
Some new treatment for an illness is being tested to see if it is better than the old way. Each of the two treatments is assigned to one half of the patients. The global results for 80 patients in the trial are given in Table 2.6. The new treatment gives a superior recovery rate.

Now consider the separate results for the two sexes, given in Table 2.7. Note, first, that combining these two subtables gives the results of Table 2.6. But we now see that the new treatment is worse than the old for each sex taken separately! This is called *Simpson's paradox*.

A treatment that seems to improve recovery from an illness, in a causal way, for the whole group, is seen to do the opposite when a new variable is introduced. This arises because males have a lower recovery rate from the disease and

Table 2.7. Separate results for the two sexes for testing a new medical treatment.

| | Female | | | |
| | Recover | | | Recovery |
Treatment	Yes	No	Total	rate
New	18	12	30	60%
Old	7	3	10	70%
	Male			
	Recover			Recovery
Treatment	Yes	No	Total	rate
New	2	8	10	20%
Old	9	21	30	30%

proportionally fewer of them received the treatment. Apparently, assignment to treatment was not at random!   □

One must draw the conclusion from this example that, although one variable may appear to depend on another, such dependence can disappear if further variables are introduced. The problem is to know which variables to use and if all such influential variables have been introduced.

Choosing individuals in a sample independently and randomly (Section 1.4.2) should reduce the chance of Simpson's paradox occurring. (Why?)

Example
Notice, in Table 2.7, that the numbers of people of each sex are not the same under each given treatment: 75% of those receiving the new treatment are women whereas only 25% of those under the old treatment are. Both these cannot simultaneously be representative of the population under study!   □

2.2   Models for binary response variables

In a small table with only two variables, each with two categories, relationships of dependence can be directly calculated by the simple odds ratio methods described above. Extending such an approach quickly becomes next to impossible as the data become more complex. Then, formal model construction can be of great help.

In our exploration of models for categorical data, we shall require methods to handle several types of additional complexity as compared with a simple 2×2 contingency table. In order of increasing difficulty, these will be

(1) one binary response and one binary explanatory variable, adequately handled by an odds ratio (Section 2.2.2);

(2) an explanatory variable with more than two categories (Section 2.2.3);

(3) more than one nominal explanatory variable (Section 2.2.4);

(4) a quantitative explanatory variable (Sections 2.2.5 and 2.3.3);

(5) a response variable with more than two categories (Section 2.3.1);

(6) several response variables (Section 2.3.2); and

(7) a response variable with ordered categories (Section 2.3.4).

In order to make clear what such more complex models can tell us, I shall begin by looking at simple cases where the results are fairly obvious even without a model.

Thus, in this section, I shall concentrate on the simplest case, where the response variable can only take on two different values; this is called a *binary response variable*. As is often the case in statistical modelling, I shall assume that the observations on different individuals are independent (see, however, Chapter 6). Then, the appropriate distribution will be the binomial of Equation (1.9). This has only one parameter, $\pi_1 = 1 - \pi_2$, which summarises the form of the distribution.

Here, we shall be interested in how this probability distribution changes in different subgroups of the population. This means that we shall be modelling *conditional* probabilities.

Example (continued)

Think of the differences in graduate plans between the two sexes in Table 1.3, where we have a 2×2 contingency table. We can label the conditional probabilities, $\pi_{1|1}$ for the probability of a male planning to do a PhD and $\pi_{1|2}$ for that of a female so planning. The second index indicates the subgroup. We are interested in studying the differences in the form of the distribution between the two subgroups; this information is contained in the parameters, $\pi_{i|j}$. These were calculated, as percentages, in Table 1.5 and were also represented as a pair of histograms in Figure 1.4. Thus, we shall be interested in studying models of the changes between these two simple histograms. □

2.2.1 MODELS BASED ON LINEAR FUNCTIONS

Many of the basic statistical models for studying dependence have the same fundamental form. The effects of the various influences are assumed to combine with each other in a simple way, yielding *linear models*.

Additive functions   One way to construct a model is to begin by assuming that there is some mean probability of the response across the whole population. Then, the individuals in the two categories of the explanatory variable will differ, in some way, from this mean.

Example (continued)

Because all the students in Table 1.3 were in the same class, we can assume that there is some common influence on all their plans to do a PhD. However, there

may also be differences between the two sexes. Let us, then, decompose each
of the two conditional probabilities of doing a PhD into a mean term and a sex-
specific term so that we have

$$\pi_{1|1} = \mu + \alpha_1$$
$$\pi_{1|2} = \mu + \alpha_2 \tag{2.3}$$

with μ the common term and α_j the specific term for each sex. □

We already know how to calculate (estimate) the values of $\pi_{i|j}$ from our data;
they are given by $\widehat{\pi_{i|j}} = n_{ij}/n_{\bullet j}$. However, if we substitute these values into the
Equations (2.3), we obtain two equations with three unknowns, $\hat{\mu}$, $\widehat{\alpha_1}$, and $\widehat{\alpha_2}$.
(Because the actual data are being used, I have placed 'hats' or circumflexes over
the parameters.)

Example (continued)
For the PhD data, the equations are

$$0.294 = \hat{\mu} + \widehat{\alpha_1}$$
$$0.545 = \hat{\mu} + \widehat{\alpha_2}$$

□

Thus, we are not able to solve the equations as they stand. We require some
additional constraint on them. However, in fact, we have not used all our assumptions. We stated that μ was the mean. If this is so, the two categories must be
situated equally on the two sides of μ. Thus, we have the additional constraint
that $\alpha_1 + \alpha_2 = 0$ or $\alpha_1 = -\alpha_2$. The equations become

$$\pi_{1|1} = \mu + \alpha_1$$
$$\pi_{1|2} = \mu - \alpha_1$$

so that we now have two equations with two unknowns. In other words, the two
probability parameters are replaced by two new parameters, μ and $\alpha_1 = -\alpha_2$. We
hope that these may be more interpretable.

This is our first example of a model based on a linear function. For short, it is
often simply called a *linear model*. It consists of a system of one or more equations that are linear in some unknown parameters. We use such models to attempt
to describe how the parameters of a probability distribution of a response variable
change when the explanatory variable changes. Linear functions are often chosen
because the equations are easy to solve. Unfortunately, in many circumstances,
the assumption of linearity may not be too realistic.

We can solve for the parameters by first adding the two equations,

$$\mu = \frac{\pi_{1|1} + \pi_{1|2}}{2}$$

This is indeed the mean conditional probability for the first category of response. Then, α_1 is the difference from this mean for males and α_2 for females:

$$\alpha_1 = \frac{\pi_{1|1} - \pi_{1|2}}{2}$$
$$= -\alpha_2$$

Example (continued)
For the PhD data, $\hat{\mu} = 0.420$, $\widehat{\alpha_1} = -0.126$, and $\widehat{\alpha_2} = 0.126$.   □

The equation $\alpha_1 + \alpha_2 = 0$ is called the *mean constraint*. Notice that this term is being applied to the constraint on a specific parameter, here α_j. It does not depend on what is on the left-hand side of Equations (2.3). Thus, it can have wide use in model construction.

If we are primarily interested in the differences between the two categories, there are other possibilities to reduce the number of unknown parameters in Equations (2.3) from three to two. We may take one category of special interest, say the first, as a baseline for comparison. Then, we compare the other with it:

$$\pi_{1|1} = \mu$$
$$\pi_{1|2} = \mu + \alpha$$

the *baseline constraint*. This is equivalent to setting $\alpha_1 = 0$ and $\alpha_2 = \alpha$ in Equations (2.3). Again, this construction only applies to the parameter α_j and not to the left-hand side of the equations.

Now, μ has a very different interpretation. It is not a mean but just the conditional probability $\pi_{1|1}$ of the first category of response for those individuals in the first category of the explanatory variable. Then, α is the difference in probability between the categories of explanatory variable:

$$\alpha = \pi_{1|1} - \pi_{1|2}$$

Although we do not have the same interpretation in terms of a common factor, we can see that $\alpha = 2\alpha_1$, so that both parameters are giving us the same type of information, comparing the distributions in the two subgroups.

Example (continued)
For the PhD data, with these constraints, we have $\hat{\mu} = 0.294$ and $\hat{\alpha} = 0.251$.   □

The models that we have constructed are based on differences in probabilities. They have a number of drawbacks that usually make them unacceptable. A fundamental one is that they decompose the probability additively. This is analogous to the addition law for mutually exclusive events.

Example (continued)
With the mean constraint, it is as if, in the model for PhD plans, we were assuming that *either* the mean factor, represented by μ, *or* the sex-specific factor,

represented by α, was influencing the probability of the yes response. This would make no sense.   □

Another limitation of such models is that they only use the first category of the response variable.

Models based on multiplicative functions   As an alternative to the additive functions above, let us now develop a model based on the product law for compound events.

As for the previous models, I shall assume that there are two influences on the response. Here, these are some mean factor in the whole population and, conditional on it, some factor specific to a given category of the explanatory variable:

$$\pi_{1|1} = \mu'\alpha_1'$$
$$\pi_{1|2} = \mu'\alpha_2'$$

Notice that, when the parameters describing the influences added together in the function, I called them 'terms' whereas, when they multiply, I call them factors. An appropriate constraint here is $\alpha_1'\alpha_2' = 1$ because the values must both be nonnegative and, if one probability is greater than μ', the other must be smaller.

Example (continued)
Again, we have two influences on the graduate plans, some mean factor and some factor specific to a given sex.   □

The model might also be constructed using a baseline approach, choosing one category of special interest:

$$\pi_{1|1} = \mu'$$
$$\pi_{1|2} = \mu'\alpha'$$

with $\alpha_1' = 1$ and $\alpha_2' = \alpha'$.

As they stand, these model functions are *nonlinear*, because they involve products of two unknown parameters. Fortunately, certain nonlinear functions can be *linearised*. Here, products can be transformed into sums by applying logarithms. Thus, these equations may also be written in a linear form:

$$\log(\pi_{1|1}) = \log(\mu') + \log(\alpha_1')$$
$$= \mu + \alpha_1$$
$$\log(\pi_{1|2}) = \log(\mu') + \log(\alpha_2')$$
$$= \mu + \alpha_2$$

where I have set $\mu = \log(\mu')$ and $\alpha_j = \log(\alpha_j')$. Logarithms must also be applied to the constraints. Then, $\log(\alpha_1'\alpha_2') = \alpha_1 + \alpha_2 = 0$, the mean constraint, or $\log(\alpha_1') = \alpha_1 = 0$ and $\alpha_2 = \alpha$, the baseline constraint.

Example (continued)
For the PhD plans, the equations become

$$\log(0.294) = \hat{\mu} + \widehat{\alpha_1}$$
$$\log(0.545) = \hat{\mu} + \widehat{\alpha_2}$$

With the mean constraints, this yields $\hat{\mu} = -0.915$ and $\widehat{\alpha_1} = -0.309 = -\widehat{\alpha_2}$, whereas, with the baseline constraint, they are $\hat{\mu} = -1.224$ and $\hat{\alpha} = 0.618$. □

By dividing the second equation of the model (without logarithms) by the first, with either constraint, we see that

$$\frac{\pi_{1|2}}{\pi_{1|1}} = \frac{\alpha'_2}{\alpha'_1}$$
$$= \alpha'$$

is the relative risk.

2.2.2 LOGISTIC MODELS

Although one fundamental problem has been eliminated by taking logarithms, both the additive probability differences and the multiplicative relative risk models still have at least two other drawbacks.

(1) The models are not symmetric in $\pi_{1|j}$ and $\pi_{2|j}$. They use only the first category of response, but another model could be constructed for the second category. We would hope that the two models would tell us the same thing about changes in distribution, but this may not be so, as we saw with the relative risk in Section 2.1.3 above.

(2) Probabilities must lie between zero and one. In the additive model, no constraint is placed on the probabilities. In the multiplicative models, the values must be greater than zero, but nothing prevents them from being greater than unity.

The solution to both these problems is to use the odds instead of the probability.

Thus, the simplest model that is acceptable according to these criteria is based on the log odds:

$$\log\left(\frac{\pi_{1|j}}{\pi_{2|j}}\right) = \log\left(\frac{\pi_{1|j}}{1-\pi_{1|j}}\right) \qquad j = 1,2 \qquad (2.4)$$
$$= \mu + \alpha_j$$

This is called a *binary logistic model*. It is a multiplicative function of the probability ratios or odds, but is linear in the log odds or logit.

Again, equivalent logistic models can be constructed using either the mean constraints, $\sum_j \alpha_j = 0$, or the baseline constraints, say $\alpha_1 = 0$. The first is most common in the literature, but the second is often the default option (or sometimes even the only possibility) in statistical software.

We immediately see that the model is symmetric in the two probabilities, taken as an odds ratio. Because of the logarithm of this ratio, the probabilities must lie in the interval, $[0, 1]$. This can be most clearly seen by back-transforming Equation (2.4) to obtain the probabilities:

$$\pi_{1|j} = \frac{e^{\mu + \alpha_j}}{1 + e^{\mu + \alpha_j}}$$

$$\pi_{2|j} = \frac{1}{1 + e^{\mu + \alpha_j}}$$

Both these values must necessarily be non-negative and it is not possible for either to be larger than one. As well, their sum is equal to one. (The reader may also like to look at the graph of a slightly more complex logistic model, given in Figure 2.5 below, which more clearly shows these limits.)

Here, I have used, for the first time, $e = 2.71828\cdots$, the base of natural logarithms.

It is also possible to solve Equation (2.4) for the new parameters. For example, with the mean constraints, we obtain

$$\mu = \frac{1}{2}\left[\log\left(\frac{\pi_{1|1}}{\pi_{2|1}}\right) + \log\left(\frac{\pi_{1|2}}{\pi_{2|2}}\right)\right]$$

$$= \frac{1}{2}\log\left(\frac{\pi_{1|1}\pi_{1|2}}{\pi_{2|1}\pi_{2|2}}\right)$$

$$\alpha_1 = \frac{1}{2}\left[\log\left(\frac{\pi_{1|1}}{\pi_{2|1}}\right) - \log\left(\frac{\pi_{1|2}}{\pi_{2|2}}\right)\right]$$

$$= \frac{1}{2}\log\left(\frac{\pi_{1|1}\pi_{2|2}}{\pi_{2|1}\pi_{1|2}}\right)$$

$$= -\alpha_2$$

The first (μ) is obtained by adding the two equations of (2.4) together and the second (α_1) by subtracting the second of these from the first. Thus, we see that, with this constraint, α_1 is just one half the log odds ratio.

The values of these parameters can be estimated directly from a contingency table, using $\widehat{\pi_{i|j}} = n_{ij}/n_{\bullet j}$:

$$\hat{\mu} = \frac{1}{2}\log\left(\frac{n_{11}n_{12}}{n_{21}n_{22}}\right)$$

$$\widehat{\alpha_1} = \frac{1}{2}\log\left(\frac{n_{11}n_{22}}{n_{21}n_{12}}\right) \tag{2.5}$$

$$= -\widehat{\alpha_2}$$

Neither of these parameter estimates involves the marginal totals; they all cancel out. This logistic model has the special characteristic that it is the only model that allows the relationship among variables in a contingency table to be modelled independently of the marginal structure. This is an additional advantage of this model over the additive probability differences and multiplicative relative risk models.

Example (continued)
For the differences in graduate plans between the sexes, in Table 1.3, the parameter values can be estimated. The two equations, based on (2.4), are

$$\log\left(\frac{5}{12}\right) = -0.875 = \hat{\mu} + \widehat{\alpha_1}$$

$$\log\left(\frac{6}{5}\right) = 0.182 = \hat{\mu} - \widehat{\alpha_1}$$

These may be solved, or Equations (2.5) applied directly, to obtain $\hat{\mu} = -0.347$ and $\widehat{\alpha_1} = -0.529$. The negative value of $\widehat{\alpha_1}$ indicates that males have a below average odds, and probability, of planning to do a PhD.

We can also back-transform these values from logarithms to probability ratios. Thus, the average odds of planning, versus not planning, to do a PhD are

$$e^{\hat{\mu}} = e^{-0.347}$$
$$= 0.707$$

Males have

$$e^{\widehat{\alpha_1}} = e^{-0.529}$$
$$= 0.589$$

times this average odds of having PhD plans, whereas the females have

$$e^{\widehat{\alpha_2}} = e^{0.529}$$
$$= 1.697$$

times.

If we multiply this average odds by the factor for the males, we obtain 0.417, which is just the odds, 5/12, of a male planning to do a PhD, and similarly for the females, $1.2 = 6/5$. □

In Section 2.1.3 above, we saw that a log odds ratio of zero indicates independence between the two variables. Thus, $\widehat{\alpha_1} = 0$ would indicate independence. For a nonzero value, the sign of $\widehat{\alpha_1}$ indicates the direction of departure from independence and the magnitude indicates the strength of dependence.

Example
Table 2.5 was constructed explicitly to have independence between the two variables, by using the fitted values from that model. We saw that the conditional

probabilities in the two subgroups were equal. Let us apply the logistic model to these imaginary data. From Equation (2.5), we obtain

$$\hat{\mu} = \frac{1}{2} \log \left(\frac{6.68 \times 4.32}{10.32 \times 6.68} \right)$$
$$= -0.435$$
$$\widehat{\alpha_1} = \frac{1}{2} \log \left(\frac{6.68 \times 6.68}{4.32 \times 10.32} \right)$$
$$= 0.000$$
$$= -\widehat{\alpha_2}$$

As expected, the log odds ratio is zero when the variables are independent. Here, in this fictitious table, the probabilities of planning to do a PhD for the two sexes are not different from the average.   □

As we saw in Section 2.1.3, the log odds is not defined if any of the four frequencies is zero. In order to obtain *approximate* estimates of the parameters, the zero frequencies can be replaced by some small value such as $1/2$.

In order to gain some feel for the interpretation of the logistic model for these simple 2×2 contingency tables, it will be useful to look at a series of similar tables.

Example
The untouchables in India are a social group who are supposed to pollute ritually any caste Hindu with whom contact is made. A study was conducted to learn about social and economic restrictions on this group. Interest centred on how laws might have eased their way of life after the independence of India from British rule. The data are observations of rural villages, the individual unit of observation, in the state of Gujarat, India. The villages were classified into four types:

IA untouchables have complete access to the village well;

IB they have access to the well, but with some restrictions;

II   there is a separate well for untouchables;

III there is no well for the untouchables in the village; they must go to the village well or water tap when Savarnas (caste Hindus) are present and ask them for water; this is the classical situation as it existed through the centuries.

Because water is so important in such a society, due both to the climate and to the religion, through ritual pollution, I shall take it to be the explanatory variable. For the moment, I shall only consider the first two village types, IA and IB, in order to have a binary explanatory variable.

A series of seven different response variables is available. These indicate various other restrictions on the untouchables in the villages, as presented in Table 2.8.

Table 2.8. Restrictions on untouchables in Gujarat villages. (Desai, 1973, pp. 26–27)

(1) Enter shop	Yes	No	Total
IA	3	2	5
IB	2	10	12
Total	5	12	17

$\hat{\mu} = -0.602, \widehat{\alpha_1} = 1.007$

(2) Give and receive in shop	Yes	No	Total
IA	3	2	5
IB	5	7	12
Total	8	9	17

$\hat{\mu} = 0.035, \widehat{\alpha_1} = 0.371$

(3) Tailor serves untouchables	Yes	No	Total
IA	2	2	4
IB	3	3	6
Total	5	5	10

$\hat{\mu} = 0.000, \widehat{\alpha_1} = 0.000$

(4) Enter temple	Yes	No	Total
IA	1	4	5
IB	1	11	12
Total	2	15	17

$\hat{\mu} = -1.892, \widehat{\alpha_1} = 0.506$

(5) Attend Panchayat meetings	Yes	No	Total
IA	3	2	5
IB	4	3	7
Total	7	5	12

$\hat{\mu} = 0.347, \widehat{\alpha_1} = 0.059$

(6) Work with Savarna in fields	Yes	No	Total
IA	4	1	5
IB	7	5	12
Total	11	6	17

$\hat{\mu} = 0.861, \widehat{\alpha_1} = 0.525$

(7) Enter Savarna house	Yes	No	Total
IA	2	3	5
IB	1	12	13
Total	3	15	18

$\hat{\mu} = -1.445, \widehat{\alpha_1} = 1.040$

(1) In villages where there are shops, untouchables can enter the shop.
(2) The shopkeeper gives things and receives money, perhaps touching the untouchable customer.
(3) In villages where there are tailors, they are not concerned with untouchability while taking measurements or while taking the cloth for stitching.
(4) Untouchables can enter the temple.
(5) Untouchable members of the Panchayat (village council) do not need to keep a distance from caste Hindu members when attending meetings.
(6) Savarna agricultural labourers and untouchable labourers work together in the fields and the former do not mind if they are touched by the latter.
(7) Untouchables can enter the private homes of Savarnas.

The parameter estimates for each logistic model are given after each table.

We see that tables can have very different social structures. Thus, as can be seen from the appropriate margin frequencies, entry to the temple (4) is rare in these villages, whereas Savarna and untouchables working together in the fields (6) is common.

However, tables with such different social structure may show similar relationships between two variables. Thus, the relationship of access to water, as given by $\widehat{\alpha_1}$, is about the same both for entry to the temple (4) and for working with the Savarna (6). The strongest relationships of dependence are found for entry to the shop (1) and to the Savarna houses (7), and the weakest for the tailor serving them (3) and for attending the Panchayat meeting (5). All relationships are non-negative, showing that less restrictions on the well go together with less restrictions on other activities.

Although I have only taken the two types of villages where access to the water is most similar, we have found some strong relationships to other restrictions. □

2.2.3   ONE POLYTOMOUS EXPLANATORY VARIABLE

The binary logistic model for data in a 2×2 contingency table provides us with no further information than that which we can obtain simply by calculating the (log) odds ratio. It is now time to explore the power of this modelling approach by seeing how it can easily be extended to more complex situations. Thus, I shall now proceed to the second level of complexity, as listed at the beginning of this Section 2.2.

Obviously, not all explanatory variables have only two categories. As we saw in Section 1.1.2, when a nominal variable has more than two categories, it is called polytomous. Thus, a first step is to extend our models in this direction. Fortunately, as it is written, Equation (2.4) is immediately applicable to $2 \times J$ contingency tables, where the explanatory variable can take $J > 2$ values. Instead of two equations to solve, we shall now have J equations. If we use a model with mean constraint, then we have $\sum_j \alpha_j = 0$. If we require a model with a baseline, say for the first category, we set $\alpha_1 = 0$. These will correspond to two different *sets* of odds ratios.

Example (continued)
Consider the complete table, for all four types of Gujarat village, for the response variable 'give and receive in the shop', as presented in Table 2.9. The corresponding changes in the histograms that we wish to study are represented in Figure 2.1. The pattern of variability is already more obvious in the histogram than in the frequency table.

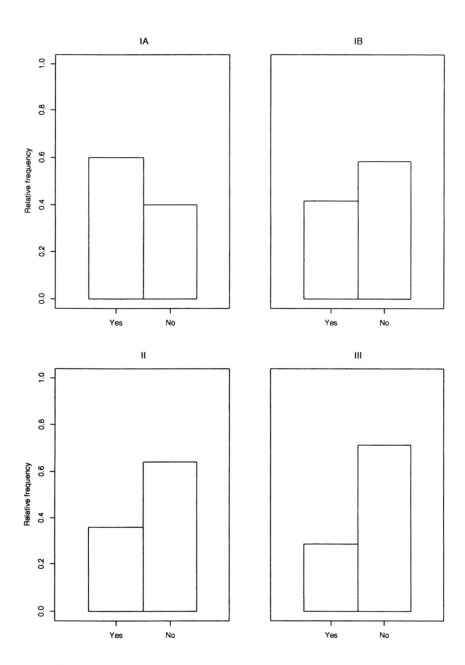

Fig. 2.1.   Histograms for giving and receiving in the shop for the four types of Indian villages.

Table 2.9. Restrictions on untouchables giving and receiving in the shops in Gujarat villages. (Desai, 1973, pp. 26–27)

(2) Give and receive in shop			
Type of village	Yes	No	Total
IA	3	2	5
IB	5	7	12
II	14	25	39
III	2	5	7
Total	24	39	63

The equations to solve are

$$\log\left(\frac{3}{2}\right) = 0.405 = \hat{\mu} + \widehat{\alpha_1}$$

$$\log\left(\frac{5}{7}\right) = -0.336 = \hat{\mu} + \widehat{\alpha_2}$$

$$\log\left(\frac{14}{25}\right) = -0.580 = \hat{\mu} + \widehat{\alpha_3}$$

$$\log\left(\frac{2}{5}\right) = -0.916 = \hat{\mu} + \widehat{\alpha_4}$$

If we use the mean constraints with Equation (2.4), we obtain $\hat{\mu} = -0.357$, $\widehat{\alpha_1} = 0.762$, $\widehat{\alpha_2} = 0.020$, $\widehat{\alpha_3} = -0.223$, and $\widehat{\alpha_4} = -0.560$. The odds of an untouchable being able to give and receive in the village shop are decreasing as the restrictions on the access to water increase.

If we use the baseline constraints, with $\alpha_1 = 0$, we have $\hat{\mu} = 0.405$, $\widehat{\alpha_2} = -0.742$, $\widehat{\alpha_3} = -0.985$, and $\widehat{\alpha_4} = -1.322$. Notice that the difference between any pair of α_j values is identical under the two constraints.

Notice also that the biggest difference between α_j values for adjacent types of villages is between IA and IB, with a value of 0.742 using either constraints, whereas the difference between types IB and III is only 0.580. Even the smallest restriction on access to water is very important.   □

2.2.4 SEVERAL EXPLANATORY VARIABLES

As we have seen, for example with Simpson's paradox, it is important to be able to include more than one variable in a model. This is the third step in the development of models for increasingly complex data. Unfortunately, the addition of a second explanatory variable adds a number of complications to the model, and to its interpretation. Nevertheless, once this jump has been made, addition of still further explanatory variables will be relatively straightforward.

Table 2.10. Sources of knowledge of cancer. (Lombard and Doering, 1947, from Potter)

	Radio			
	Yes		No	
	Knowledge			
Lectures	Good	Poor	Good	Poor
Yes	36	18	39	42
No	166	192	427	809

Two explanatory variables When two explanatory variables are being used in a model to describe changes in response, each of them will subdivide the population in such a way that the distribution of the response variable can be different, as with the two-way table. But, in addition, the relationship of the response distribution to the first explanatory variable may vary depending on what value the second explanatory variable takes, and vice versa. This is called an *interaction* between the two explanatory variables with respect to the response variable. This is a precise technical definition of interaction that should not be confused with its everyday use.

Example
Table 2.10 gives data about whether people have good or poor knowledge concerning cancer as related to two of the many sources from which it might be acquired: radio and newspapers. The histograms given in Figure 2.2 illustrate the differences in probability of having knowledge about cancer.

The effect of radio may be different for those people also attending lectures from that for those who do not. For example, lectures may aid people in understanding what they hear on the radio. Inspect the histograms to see if this might be true. □

With two explanatory variables, there will now be four sets of parameters, the mean μ, the main effect of the first explanatory variable α_j, the main effect of the second explanatory variable β_k, and the interaction between them γ_{jk}. Each set may contain one or more parameters, depending on the number of categories of the corresponding explanatory variable.

Then, we can write the logistic model as

$$\log\left(\frac{\pi_{1|jk}}{\pi_{2|jk}}\right) = \mu + \alpha_j + \beta_k + \gamma_{jk} \qquad j = 1,\ldots,J, \quad k = 1,\ldots,K \quad (2.6)$$

If we work with a model about the mean, we have the constraints, $\sum_j \alpha_j = 0$, $\sum_k \beta_k = 0$, $\sum_j \gamma_{jk} = 0$ for all k, and $\sum_k \gamma_{jk} = 0$ for all j. The appropriate baseline constraints could also be used, but they become more difficult to interpret as interactions are added to the model. I shall not use them further in this chapter.

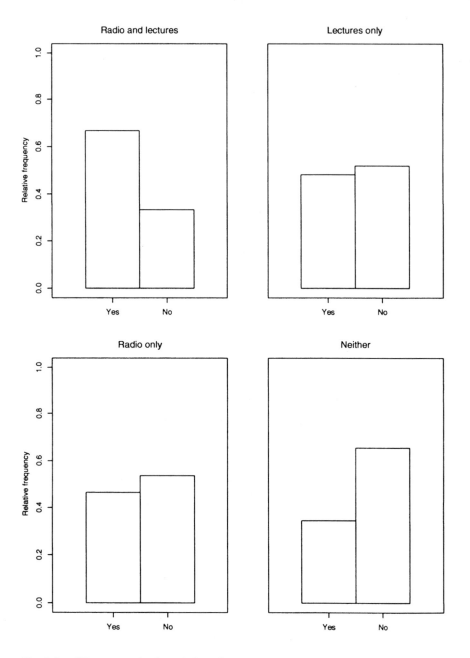

Fig. 2.2. Histograms for knowledge of cancer as related to listening to the radio and attending lectures.

Thus, for a $2 \times 2 \times 2$ table, we shall now have four equations to solve, whereas a $2 \times 2 \times 3$ table results in six equations, and so on.

Example (continued)
For the data on knowledge of cancer, let j index radio and k lectures. As might be expected, we see from the histograms given in Figure 2.2 that both sources of knowledge about cancer have positive effects.

The four equations to solve to obtain the parameter estimates are

$$\log\left(\frac{36}{18}\right) = 0.693 = \hat{\mu} + \widehat{\alpha_1} + \widehat{\beta_1} + \widehat{\gamma_{11}}$$

$$\log\left(\frac{166}{192}\right) = -0.146 = \hat{\mu} + \widehat{\alpha_1} - \widehat{\beta_1} - \widehat{\gamma_{11}}$$

$$\log\left(\frac{39}{42}\right) = -0.074 = \hat{\mu} - \widehat{\alpha_1} + \widehat{\beta_1} - \widehat{\gamma_{11}}$$

$$\log\left(\frac{427}{809}\right) = -0.639 = \hat{\mu} - \widehat{\alpha_1} - \widehat{\beta_1} + \widehat{\gamma_{11}}$$

where α_j is the main effect of radio, β_k that of lectures, and γ_{jk} their interaction. If we add all four equations together and divide by four, we obtain $\hat{\mu} = -0.041$. If we add the first two equations, divide by two, and subtract the value just obtained for $\hat{\mu}$, we have $\widehat{\alpha_1} = 0.315 = -\widehat{\alpha_2}$. If we perform the same steps with the first and third equations, we get $\widehat{\beta_1} = 0.351 = -\widehat{\beta_2}$, whereas with the first and last equations, we have $\widehat{\gamma_{11}} = 0.068 = -\widehat{\gamma_{12}} = -\widehat{\gamma_{21}} = \widehat{\gamma_{22}}$.

We see that attending lectures has a slightly larger positive effect on cancer knowledge than listening to the radio. This can be seen clearly in the histograms of Figure 2.2. The interaction is positive but close to zero so that the effects of the two sources are cumulative, with a small positive reinforcement. □

Once we begin to manipulate more complex models, we shall generally wish to simplify them as much as possible. We shall have to wait until Chapter 3 for criteria to do this. However, it is important already to understand that there are certain restrictions on what may be valid simplifications. In the previous example, the interaction was considerably smaller than the two main effects; eliminating it may be a useful simplification to investigate. However, it does not always make sense to eliminate a parameter even if it has an extremely small value.

Example
Lazarsfeld (1955) had particular experience in studying radio listening habits during the Second World War to aid the USA propaganda machine in reaching its audience. Of the several types of radio programmes that he studied, I shall here look at listening to classical music (indexed by i) as it depends on the two explanatory variables, age (indexed by j) and education (indexed by k). The education variable separates those who completed secondary school from those who did not,

Table 2.11. Classification of classical music listeners by age and education. (Lazarsfeld, 1955, pp. 115–125)

	Education			
	High		Low	
	Listen to classical music			
Age	Yes	No	Yes	No
Old	210	190	170	730
Young	194	406	110	290

whereas ages are more or less than 40 years old. The results are shown in Table 2.11. Lazarsfeld states that these data represent, in somewhat stylised form, the results obtained in many studies of radio listening tastes.

The histograms illustrating the differences in probability of listening to classical music are shown in Figure 2.3. The four equations to solve to obtain the parameter estimates are

$$\log\left(\frac{210}{190}\right) = 0.100 = \hat{\mu} + \widehat{\alpha_1} + \widehat{\beta_1} + \widehat{\gamma_{11}}$$

$$\log\left(\frac{170}{730}\right) = -1.457 = \hat{\mu} + \widehat{\alpha_1} - \widehat{\beta_1} - \widehat{\gamma_{11}}$$

$$\log\left(\frac{194}{406}\right) = -0.738 = \hat{\mu} - \widehat{\alpha_1} + \widehat{\beta_1} - \widehat{\gamma_{11}}$$

$$\log\left(\frac{110}{290}\right) = -0.969 = \hat{\mu} - \widehat{\alpha_1} - \widehat{\beta_1} + \widehat{\gamma_{11}}$$

where α_j is the main effect of age, β_k that of education, and γ_{jk} their interaction. Solving these equations in the same way as we did for the first example, we obtain $\hat{\mu} = -0.766$, $\widehat{\alpha_1} = 0.088 = -\widehat{\alpha_2}$, $\widehat{\beta_1} = 0.447 = -\widehat{\beta_2}$, and $\widehat{\gamma_{11}} = 0.332 = -\widehat{\gamma_{12}} = -\widehat{\gamma_{21}} = \widehat{\gamma_{22}}$.

From the main effects, the distribution of listeners to classical music changes considerably, on average, with differences in education; higher education yields more listeners. On the other hand, the differences in age, on average across education levels, shows virtually no difference in the response distribution.

However, the interaction effect indicates that difference with education is much greater among the old than among the young: for the old people, we have $\widehat{\beta_1} + \widehat{\gamma_{11}} = 0.779$, whereas, for the young, $\widehat{\beta_1} + \widehat{\gamma_{21}} = 0.115$. Among the old, listening to classical music depends strongly on education, whereas, for the young, it does not. We can look at this another way: at the high level of education, the difference for age is $\widehat{\alpha_1} + \widehat{\gamma_{11}} = 0.420$ whereas, at the low level, it is $\widehat{\alpha_1} + \widehat{\gamma_{12}} = -0.244$. At the high level of education, old people listen proportionally more to classical music than young people, but, at the low level, young people listen more.

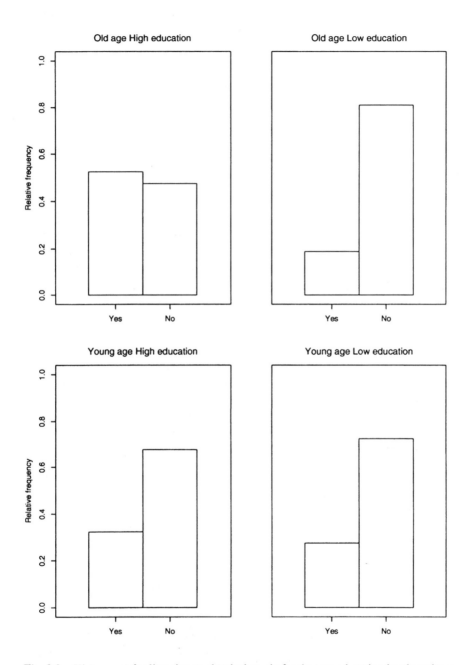

Fig. 2.3. Histograms for listening to classical music for the two education levels and two ages.

Table 2.12. Classification of classical music listeners separately by age and by education. (Lazarsfeld, 1955, pp. 115–125)

Age	Listen to classical music	
	Yes	No
Old	380	920
Young	304	696

Education	Listen to classical music	
	Yes	No
High	404	596
Low	280	1020

Suppose, now, that, instead of applying Equation (2.6) to all the data, we apply Equation (2.4) separately for old and for young people. For the old people, we obtain $\hat{\mu} = -0.679$ and $\widehat{\alpha_1} = 0.779$, whereas, for the young, we have $\hat{\mu} = -0.854$ and $\widehat{\alpha_1} = 0.115$. The α_{ij} values are those we just obtained above by combining parameters $(\hat{\beta}_1 + \widehat{\gamma_{j1}})$ in the more complex model; the μ values could be obtained by combining the remaining parameters in the same way. We could also perform the same operations separately for the two education levels, again reproducing the other results obtained above. Thus, Equation (2.6) simultaneously gives all the results obtainable by applying Equation (2.4) four times to the data.

We may ask what would happen if we applied Equation (2.4) to only one of the variables, education or age, ignoring the other. The two-way marginal tables are given in Table 2.12, obtained by adding the appropriate values in Table 2.11. For education, we obtain $\hat{\mu} = -0.841$ and $\widehat{\alpha_1} = 0.452$, whereas for age, we have $\hat{\mu} = -0.856$ and $\widehat{\alpha_1} = -0.028$. The values of $\widehat{\alpha_1}$ are similar to the main effect parameters of the two variable model. By only fitting these two simple models, we would conclude that listening to classical music varies little with age, missing the important interaction with education brought out by the two variable model. This is a mild form of Simpson's paradox.   □

In this example, it does not make sense to eliminate the parameter for age even though it is much smaller than the others. This is because the interaction is required in the model. Thus, it is wise rarely to eliminate a simple term in a model (such as that for age) if a more complex term containing it is retained (the interaction between age and education). In other words, we should retain the *hierarchical* nature of a model.

Three explanatory variables As we add more explanatory variables to the model, the number of possible interaction terms necessary increases quickly. For three explanatory variables, we have

$$\log \left(\frac{\pi_{1|jkl}}{\pi_{2|jkl}} \right) = \mu + \alpha_j + \beta_k + \delta_l + \gamma_{1jk} + \gamma_{2jl} + \gamma_{3kl} + \gamma_{4jkl}$$

Table 2.13. Sources of knowledge of cancer. (Lombard and Doering, 1947, from Potter)

		Radio			
		Yes		No	
		Knowledge			
Newspaper	Lectures	Good	Poor	Good	Poor
Yes	Yes	31	12	34	24
	No	137	126	276	333
No	Yes	5	6	5	18
	No	29	66	151	476

If all variables are binary, a $2 \times 2 \times 2 \times 2$ table, there will be eight equations to solve. Generally, we shall hope that as many higher order interactions as possible are close to zero, because they are often difficult to interpret.

Example (continued)
In Table 2.10, we examined data concerning how knowledge of cancer depended on attending lectures and listening to the radio. Another source of knowledge studied was newspapers. Table 2.13 gives the data for all three of these sources of knowledge simultaneously. As before, let j index radio and k lectures, but now l will refer to newspapers. The histograms illustrating the differences in probability of having knowledge about cancer are given in Figure 2.4.

When we solve the eight equations, we obtain $\hat{\mu} = -0.280$, $\widehat{\alpha_1} = 0.287$, $\hat{\beta}_1 = 0.239$, $\hat{\delta}_1 = 0.578$, $\widehat{\gamma_{111}} = 0.138$, $\widehat{\gamma_{211}} = -0.069$, $\widehat{\gamma_{311}} = 0.112$, and $\widehat{\gamma_{4111}} = -0.055$. All four interactions are relatively small as compared with the three main effects. Newspapers are about twice as strongly associated with good knowledge of cancer as radio or lectures.

As compared with the previous results for radio and lectures alone, the effects of these two sources decrease and their interaction increases when the effect of newspapers is included in the model. This is another mild form of Simpson's paradox.   □

2.2.5 LOGISTIC REGRESSION

In Section 2.2.3, I looked at models involving explanatory variables having several categories. Suppose now that the values attached to these categories are quantitative. In such cases, we can often simplify the model very considerably by assuming a simple linear relationship among the log odds:

$$\log\left(\frac{\pi_{1|j}}{\pi_{2|j}}\right) = \beta_0 + \beta_1 x_j \tag{2.7}$$

where x_j is usually the centre of the interval for the value of the jth category of the explanatory variable. This is known as a *logistic regression model*. It is the fourth

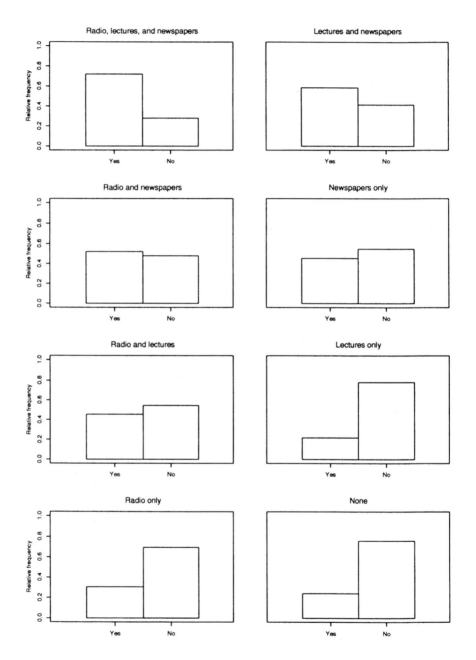

Fig. 2.4. Histograms for knowledge of cancer as related to listening to the radio, attending lectures, and reading newspapers.

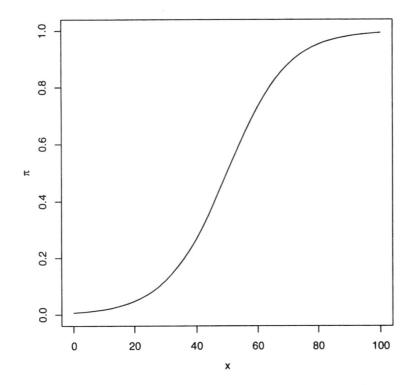

Fig. 2.5. Logistic regression model with a quantitative variable and $\beta_1 > 0$.

step in our search for models to handle increasingly complex situations. Here, the model results in a simplification but the procedure for obtaining estimates of the parameters is more difficult to implement.

As with previous binary logistic models, such as Equation (2.4), this equation can be solved for the probability,

$$\pi_{1|j} = \frac{e^{\beta_0 + \beta_1 x_j}}{1 + e^{\beta_0 + \beta_1 x_j}}$$

$$= \frac{1}{1 + e^{-\beta_0 - \beta_1 x_j}}$$

(2.8)

If we know the values of β_0 and β_1, we can then plot this *logistic curve* using a selection of values for x_j. One example of such a curve is given in Figure 2.5. We see how the probabilities can never leave the interval $[0, 1]$. As well, as x_j increases, the probabilities must all either increase, if $\beta_1 > 0$, as in the graph, or decrease, if $\beta_1 < 0$. They cannot increase and then decrease, or vice versa. Such a curve is said to be *monotone*.

In such a model, there are only two unknown parameters, instead of the J parameters if we used the logistic model in Section 2.2.3. Thus, we have here a considerably simpler model, an advantage if it describes the data well.

However, there will still be J equations, one for each category of the explanatory variable, with only two unknowns. Thus, we shall not be able to solve the equations in the simple way that we have previously used. An approximate, but usually fairly accurate, method requires the calculation of several quantities. First, let

$$z_j = \log\left(\frac{n_{1j}}{n_{2j}}\right)$$

$$w_j = \frac{n_{1j}n_{2j}}{n_{\bullet j}}$$

Then, calculate

$$A = \sum_j w_j$$

$$B = \sum_j w_j z_j$$

$$C = \sum_j w_j x_j$$

$$D = \sum_j w_j x_j^2$$

$$E = \sum_j w_j x_j z_j$$

From these, we obtain

$$\widehat{\beta_1} \doteq \frac{E - B \times C/A}{D - C^2/A}$$

$$\widehat{\beta_0} \doteq \frac{B}{A} - \widehat{\beta_1}\frac{C}{A}$$

Example
Alcohol consumption by a pregnant mother is thought to be dangerous for the child to be born. Table 2.14 gives data on daily alcohol consumption and the corresponding number of malformed children from a prospective study of maternal drinking and congenital malformations. After the first three months of pregnancy, the women were asked to complete a questionnaire on their alcohol consumption during that period. Then, following the birth of the child, various types of information were recorded on the pregnancy outcome, including the presence or absence of congenital sex organ malformations, as shown in the table. Notice that 0.28% of mothers with zero alcohol consumption have malformed children, whereas 0.26% of those with < 1 drinks per day do.

Table 2.14. Mother's alcohol consumption (average drinks per day) and birth defects. (Graubard and Korn, 1987)

Malformation	Alcohol consumption				
	0	< 1	1–2	3–5	≥ 6
Absent	17066	14464	788	126	37
Present	48	38	5	1	1

Table 2.15. Relative frequencies and logistic regression estimates for the malformation data of Table 2.14.

Malformation	Alcohol consumption				
	0	< 1	1–2	3–5	≥ 6
Relative frequency	0.9972	0.9974	0.9937	0.9921	0.9737
Logistic regression	0.9974	0.9970	0.9958	0.9906	0.9756

One problem is that we do not have exact values of alcohol consumption, but only categories when nonzero. If we take the values of x_j to be $(0, 0.5, 1.5, 4, 7)$, the estimates using the approximate method just described are $\widehat{\beta}_0 \doteq 5.953$ and $\widehat{\beta}_1 \doteq -0.324$. The negative value of the slope parameter $\widehat{\beta}_1$ shows that the probability of absence of malformation decreases with increasing alcohol consumption.

In Table 2.15, the relative frequencies can be compared with the values obtained from the logistic regression, and can be seen to be rather close. These are plotted in Figure 2.6. Because the estimated slope is negative, this curve goes down instead of rising, as in Figure 2.5. Notice, as well, that only probability values in the range $(0.96, 1)$ have been plotted, so that the S-shape is not evident.

Because of the monotonicity of the logistic curve, the model has imposed an increase in malformations for mothers with < 1 drinks per day as compared with those taking no drinks, whereas the proportion calculated above indicated the reverse. Our model, if plausible, is indicating that this may be a random fluctuation in the observed data. We shall have to wait until Section 3.5 to see how to determine whether or not the logistic regression model is plausible for these data. □

Exact estimates can be obtained by repeating the calculations several times, called *iterating*, using new values of the weights, given, each time, by $w_j = n_{\bullet j}\widetilde{\pi}_{1|j}\widetilde{\pi}_{2|j}$, where $\widetilde{\pi}_{i|j}$ are values obtained from Equation (2.8) in the previous step. This is what computer software does.

Example
The exact method, using computer software, yields the estimates $\widehat{\beta}_0 = 5.961$ and $\widehat{\beta}_1 = -0.317$, not very different from the approximate values given above. □

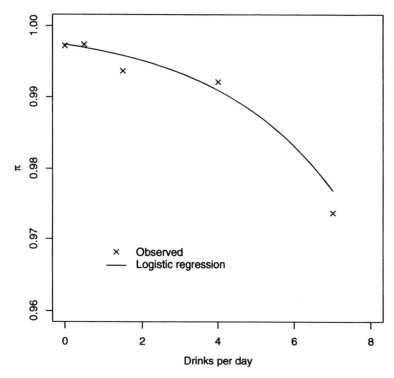

Fig. 2.6. Logistic regression for the probability of absence of birth defects, as related to the mother's alcohol consumption. Notice the limits on the y-axis.

Indicator variables Although logistic regression appears to be a special case of the other logistic models above, it is often more useful to think of these latter models as *special* cases of logistic regression. Consider Equation (2.4) for a binary explanatory variable:

$$\log\left(\frac{\pi_{1|j}}{\pi_{2|j}}\right) = \mu + \alpha_j \qquad j = 1, 2$$

This is identical to

$$\log\left(\frac{\pi_{1|j}}{\pi_{2|j}}\right) = \beta_0 + \beta_1 x_j \qquad j = 1, 2$$

if x_j is coded $(0,1)$ for a baseline constraint or $(-1,1)$ for a mean constraint. Such an explanatory variable is called an *indicator* or *dummy variable*. This procedure can easily be extended to explanatory variables with more than two categories, but then will require several indicator variables (Section 5.2.2).

2.3 Polytomous response variables

Up until now, I have only constructed models for data where there is one binary response variable. However, in many realistic situations, the response will have more than two categories.

2.3.1 POLYTOMOUS LOGISTIC MODELS

We are now ready to look at the case where the response variable is nominal, taking more than two values, say with I categories. This will be our fifth step in models of increasing complexity. Thus, we shall be modelling changes in the shape of the more general multinomial distribution instead of the special case of a binomial distribution. The generalisation of our simple binary logistic model, Equation (2.4) when there was one explanatory variable, is not as obvious as when an explanatory variable took several values in Section 2.2.3.

Comparing response categories With two response categories, only one comparison is necessary, the odds: $\pi_{1|j}/\pi_{2|j}$ (or its reciprocal). With several categories, more such pairwise comparisons are possible. For example, with three possible response values and one explanatory variable, we can compare

$$\frac{\pi_{1|j}}{\pi_{2|j}}, \quad \frac{\pi_{1|j}}{\pi_{3|j}}, \quad \frac{\pi_{2|j}}{\pi_{3|j}}$$

Unfortunately, the number of possible comparisons escalates rapidly with the number of categories. For example, with four categories, there are six pairwise comparisons and, with five categories, ten comparisons.

To facilitate examination of such relationships, we can compare all probabilities with some common value. If we recall the constraints that we applied to the parameters for explanatory variables in binary logistic models, we have at least two logical choices: a mean or a baseline category.

Consider first the latter. There will be $I - 1$ comparisons, say of the probability of the first category of response to that of each other category. This is considerably fewer than the total number of possible pairwise comparisons, especially when the response has many categories. However, we know that use of baseline categories, when there are several explanatory variables in the model, makes interpretation of interactions difficult.

Thus, let us concentrate on comparisons with a mean. Because we are working with logarithms, it makes sense to take means of the logarithms of the probabilities for a given category j of the explanatory variable:

$$\log(\mathring{\pi}_j) = \frac{1}{I} \sum_i \log(\pi_{i|j})$$

Here, I have denoted by $\mathring{\pi}_j$ the value of the probability obtained in this way for the jth category of the explanatory variable. This value is called the *geometric mean*. Back-transformation, eliminating the logarithm, shows its definition:

$$\dot{\pi}_j = \left(\prod_i \pi_{i|j} \right)^{\frac{1}{I}}$$

Notice that, for $I = 2$ in binary logistic models, $\dot{\pi}_j = \sqrt{\pi_{1|j}\pi_{2|j}}$.
    Now all comparisons may be made with this geometric mean:

$$\frac{\pi_{1|j}}{\dot{\pi}_j}, \quad \frac{\pi_{2|j}}{\dot{\pi}_j}, \quad \frac{\pi_{3|j}}{\dot{\pi}_j}$$

If we multiply all these values together, we discover that their product is equal
to one. Thus, there are really only $I - 1$ different comparisons, as for a baseline
model. Taking the ratio of any pair of them gives the corresponding odds and
subtracting their logarithms,

$$\log\left(\frac{\pi_{i|j}}{\dot{\pi}_j}\right) - \log\left(\frac{\pi_{k|j}}{\dot{\pi}_j}\right) = \log\left(\frac{\pi_{i|j}}{\pi_{k|j}}\right)$$

yields the log odds.
    When $I = 2$, we find that

$$\log\left(\frac{\pi_{1|j}}{\dot{\pi}_j}\right) = \log\left(\frac{\pi_{1|j}}{\sqrt{\pi_{1|j}\pi_{2|j}}}\right)$$
$$= \frac{1}{2}\log\left(\frac{\pi_{1|j}}{\pi_{2|j}}\right)$$

Except for the factor of $1/2$, this is the logit that was the basis for the binary
logistic model in Equations (2.4) and (2.6). Thus, we have found a suitable basis
for generalisation of binary logistic models to the case where the response has
more than two categories.

One explanatory variable   Thus, this construction in terms of geometric means
will now form the basis for *polytomous logistic models*. For one explanatory
variable, such a model can be written

$$\log\left(\frac{\pi_{i|j}}{\dot{\pi}_j}\right) = \mu_i + \alpha_{ij} \tag{2.9}$$

with one such equation for each category j of the explanatory variable, as well
as for each category of response i. This is a set of $I \times J$ linear equations to
describe how the multinomial distribution is changing in different categories of
the explanatory variable. Fortunately, we do not have to solve them all at once.
    Because the response categories are compared with a mean, we shall also use
mean constraints for the parameters: $\sum_i \mu_i = 0$, $\sum_i \alpha_{ij} = 0$ for all j, and $\sum_j \alpha_{ij} = 0$
for all i.

Table 2.16. Choice of school programme as related to social class. (Lindsey, 1981)

| Social class | School programme | | | Total |
	Academic	Vocational	General	
Capitalist	108	7	21	136
Professional	392	49	115	556
Small proprietor	288	49	126	463
White collar	197	52	94	343
Manual worker	532	269	646	1447
Total	1517	426	1002	2945

The estimates may be obtained by solving I sets of J equations. With $\widehat{\pi_{i|j}} = n_{ij}/n_{\bullet j}$, we have

$$\log(\widehat{\pi_{i|j}}) - \frac{1}{I}\sum_i \log(\widehat{\pi_{i|j}}) = \log(n_{ij}) - \frac{1}{I}\sum_i \log(n_{ij})$$

In the same way as for the binary logistic models, the denominators cancel, so that we can work directly with the frequencies in the table.

Example
The International Association for the Evaluation of Educational Achievement conducted a comparative study of primary education in a number of countries in 1964. Here, I shall look at some of the results for England. Table 2.16 gives a contingency table relating choice of school programme for 13-year-old children in England to their parents' social class. The response variable has three categories. The distributions for the five social classes, which we wish to model, are represented as histograms in Figure 2.7.

There are three sets of five equations to solve. It will be useful first to calculate the log geometric mean of the frequencies for each row. These are 3.224 (= $[\log(108) + \log(7) + \log(21)]/3$), 4.869, 4.797, 4.593, and 6.114, respectively. Now, the set of equations for the first column is

$$\log(108) - 3.224 = \widehat{\mu_1} + \widehat{\alpha_{11}}$$
$$\log(392) - 4.869 = \widehat{\mu_1} + \widehat{\alpha_{12}}$$
$$\log(288) - 4.797 = \widehat{\mu_1} + \widehat{\alpha_{13}}$$
$$\log(197) - 4.593 = \widehat{\mu_1} + \widehat{\alpha_{14}}$$
$$\log(532) - 6.114 = \widehat{\mu_1} + \widehat{\alpha_{15}}$$

Similar sets must be written down for the other two columns. Each set of five equations can be solved in exactly the same way as for the logistic models for binary response in Section 2.2.2. When the three sets of equations are solved, we obtain the values in Table 2.17. A positive estimate indicates over-representation

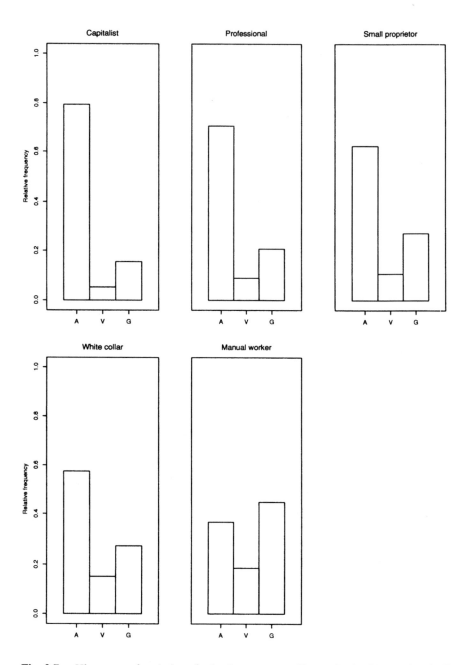

Fig. 2.7. Histograms for choice of school programme (A: academic, V: vocational, G: general) for the five social classes.

Table 2.17. Parameter estimates for the data of Table 2.16.

$\widehat{\mu}_i$	0.856	−0.864	0.009
$\widehat{\alpha}_{i1}$	0.602	−0.414	−0.188
$\widehat{\alpha}_{i2}$	0.246	−0.113	−0.133
$\widehat{\alpha}_{i3}$	0.010	−0.041	0.031
$\widehat{\alpha}_{i4}$	−0.165	0.223	−0.058
$\widehat{\alpha}_{i5}$	−0.693	0.345	0.348

and a negative value under-representation of the social class in that school pro-gramme, because values are differences from the mean.

Children of the two dominant classes are more often than average found in the academic programme, whereas those of the manual working class are more often in the vocational and general. This can be seen from the large positive values in the corresponding positions of Table 2.17. In the same way, we see that children of white collar workers are more often than average in the vocational pro-gramme, whereas those of small proprietors are about average in all programmes. Similar statements can be made about the negative values, which indicate under-representation.

From this model, the average odds of a student being in academic versus gen-eral can be calculated as $e^{0.856-0.009} = 2.33$. However, for the capitalist class, the odds are $e^{0.602-(-0.188)} = 2.20$ times this average, whereas for the manual work-ing class they are estimated to be $e^{-0.693-0.348} = 0.35$ times this average. The important thing is the difference from the average, not the average itself: the table of α_{ij} values is of most interest. These summarise the log odds ratios in a clearly interpretable way. Thus, the odds of a manual working class child being in the academic versus the general programme are

$$\frac{532/1447}{646/1447} = 0.82$$
$$= 2.33 \times 0.35$$

This value of 0.82 does not seem unreasonable until we see that it is only about one-third of the average for all social classes.   □

It may be useful to present a second, more algorithmic, procedure for ob-taining the parameter estimates in an $I \times J$ table. The following steps should be followed:

(1) From the original frequency table, construct a new table by taking the log-arithms of all frequencies.
(2) Calculate the overall mean of these values and record it at the lower right corner of the table.
(3) Calculate each column mean, subtract the overall mean, and record it below the corresponding column.
(4) Do the same for the rows, placing the values to the right.

Table 2.18. Algorithmic calculations of parameter values for the data of Table 2.16.

4.682	1.946	3.045	-1.495
5.971	3.892	4.745	0.150
5.663	3.892	4.836	0.078
5.283	3.951	4.543	-0.127
6.277	5.595	6.471	1.395
0.856	-0.864	0.009	4.719

(5) Finally, construct a second table by taking each log frequency in the body of the first table, subtracting the appropriate value recorded to the right of its row, the value below its column, and the overall mean.

The bottom margin of the first table will give the estimates of the μ_i parameters and the second table will give the α_{ij} values.

Example (continued)
For the data on schooling in England, the first table of calculations is presented in Table 2.18. In the bottom margin, we find the estimates of the μ_i parameters that we had previously. The second table, containing the $\widehat{\alpha_{ij}}$ values, is the same as Table 2.17 (except that the former does not contain the first line).   □

One interesting aspect of this algorithm is that it is completely symmetric in the two variables, not distinguishing between response and explanatory variables. Thus, these models retain this desirable characteristic of odds ratios discussed in Section 2.1.4. I shall use this feature in Section 2.3.2.

Example (continued)
The right column of Table 2.18 gives the mean parameters for the explanatory variable, social class.   □

Several explanatory variables   In the same way as for binary logistic models, the polytomous logistic model is directly extensible to several explanatory variables. For two variables, it is

$$\log\left(\frac{\pi_{i|jk}}{\pi_{jk}}\right) = \mu_i + \alpha_{ij} + \beta_{ik} + \gamma_{ijk} \tag{2.10}$$

Obviously, the calculations quickly become very onerous without the aid of a computer.

Example
Breslow and Day (1982, p. 155) present data from the Ille-et-Vilaine, France, case-control study of œsophageal cancer. Cases were 200 males diagnosed with

Table 2.19. Results of a case-control study of the dependence of œsophageal cancer on the risk factors, alcohol and tobacco consumption, in the Ille-et-Vilaine, France. (Breslow and Day, 1982, p. 155)

Alcohol	Tobacco			
	0–9	10–19	20–29	30+
	Cases			
0–39	9	10	5	5
40–79	34	17	15	9
80–119	19	19	6	7
120+	16	12	7	10
	Controls			
0–39	252	74	35	23
40–79	145	68	47	20
80–119	42	30	10	5
120+	8	6	5	3

Table 2.20. Parameter values for the separate dependence of œsophageal cancer on alcohol and tobacco consumption.

	Alcohol consumption			
	0–39	40–79	80–119	120+
Case	−0.827	−0.194	0.198	0.823
Control	0.827	0.194	−0.198	−0.823

	Tobacco consumption			
	0–9	10–19	20–29	30+
Case	−0.318	−0.005	0.016	0.306
Control	0.318	0.005	−0.016	−0.306

this cancer in a regional hospital between January 1972 and April 1974. Controls were a sample drawn from the electoral list. Among other variables, information was collected on alcohol and tobacco consumption (g/day), with results as shown in Table 2.19.

As we have seen in Section 2.1.4, case-control studies can be analysed using odds ratios. Because all logistic models are based on such ratios, they will be appropriate. Although these data can be analysed using a binary logistic model, I shall present the results when they have been calculated using a polytomous logistic model. I shall first look at the separate models, using Equation (2.9), for each explanatory variable by itself, given in Table 2.20.

Recall that negative estimates indicate under-consumption of alcohol or to-bacco whereas positive values support over-consumption of these products, with

Table 2.21. Parameter values for the simultaneous dependence of œsophageal cancer on alcohol and tobacco consumption.

	Alcohol consumption			
	0–39	40–79	80–119	120+
Case	−0.798	−0.189	0.198	0.789
Control	0.798	0.189	−0.198	−0.789

	Tobacco consumption			
	0–9	10–19	20–29	30+
Case	−0.226	−0.033	−0.009	0.268
Control	0.226	0.033	0.009	−0.268

respect to the mean. We see the clear indication that the cases consume more both of alcohol and of tobacco, as shown by the large positive $\widehat{\alpha}_{ij}$ values in the table. However, the differences for alcohol consumption are much larger than those for tobacco consumption. This may not be surprising because alcohol normally passes through the œsophagus whereas tobacco smoke does not.

For these data, it turns out that the interaction parameters are not necessary in a model containing both types of consumption simultaneously as risk factors. The parameters for this model, Equation (2.10) with $\gamma_{ijk} = 0$ for all i, j, k, are given in Table 2.21.

If we compare Tables 2.20 and 2.21, we see that the parameters for the dependence of cancer on alcohol consumption change relatively little whether tobacco consumption is in the model or not. However, those for the dependence on tobacco consumption are quite different in the two models. When alcohol consumption is added to a model where cancer depends only on tobacco consumption, the latter parameters are reduced considerably in size. Thus, some of the observed dependence of cancer on tobacco consumption can be explained by differences in alcohol consumption, whereas the reverse is not true.   □

2.3.2   LOG LINEAR MODELS

In the categorical data models that we have constructed so far, we single out one variable as the response of interest. Then, we study changes in the conditional distribution of the response within different subgroups of the population. Construction and study of models based on conditional distributions is often appropriate for at least two reasons:

(1) scientific interest centres on how the *probability distribution* of the response variable depends on, or changes with, values of the explanatory variable(s);

(2) the explanatory variable(s) can be taken to have *fixed values*, so that any randomness in them need not be modelled.

In some situations, however, no one variable can be distinguished in this way as the response. Models to handle this problem will constitute the sixth step in complexity.

Multivariate models   There are two possibilities when no one response variable is distinguishable:

(1) all variables are taken to be on the same conceptual level, with no explanatory variables;

(2) two or more variables are simultaneously responses, conditional on all others.

In the first case, we shall be interested in the joint or *multivariate* distribution of the response variables rather than a conditional distribution. In the second case, we also require a joint distribution of the responses, but it will be conditional on the explanatory variables. In both cases, we require multivariate response models. We can construct such models by using a decomposition into factors in a way similar to that for the conditional distribution for one response. The result is called a *log linear model* for categorical data.

When the distribution of a response variable is conditional on one or more explanatory variables, we often say that the response *depends* on those variables. When several response variables are inter-related in a symmetrical fashion, we can say that they are *associated*.

Let us consider first the simplest case: two response variables with no explanatory variables. We can construct a model in a way similar to what we did for a polytomous response variable. For the joint distribution of the two variables, we can write

$$\log\left(\frac{\pi_{ij}}{\dot{\pi}}\right) = \mu_i + \phi_j + \alpha_{ij} \tag{2.11}$$

with the usual mean constraints. Notice that we now require two sets of mean parameters, one for each variable. Here, $\dot{\pi}$ has no index. It is the geometric mean over the whole table, instead of for each category of the explanatory variable(s). The set of parameters α_{ij} measures the association between the two variables.

For three variables, the log linear model corresponding to the polytomous logistic model of Equation (2.10) will be

$$\log\left(\frac{\pi_{ijk}}{\dot{\pi}}\right) = \mu_i + \phi_j + \psi_k + \alpha_{ij} + \beta_{ik} + \delta_{jk} + \gamma_{ijk} \tag{2.12}$$

with suitable constraints. The four sets of parameters α_{ij}, β_{ik}, δ_{jk}, and γ_{ijk} describe all possible relationships of association among the three variables.

Because of the symmetry among the variables in log linear models, two or more variables may simultaneously be interpreted as responses, conditional on the remaining explanatory variables, whenever this is appropriate. Then, the 'interaction' parameters among the response variables describe their association.

Example
In Equation (2.12), suppose that i and j index response variables and k an explanatory variable. Then, α_{ij} measures the association between the two responses, whereas β_{ik}, δ_{jk}, and γ_{ijk} measure their dependence on the explanatory variable.□

However, we may also only interpret one variable as response, and ignore the unnecessary parameters. In this way, we obtain binary and polytomous logistic models as special cases of log linear models.

Example
In Equation (2.12), if the only response is indexed by i, we can ignore δ_{jk}. The result is the polytomous logistic model of Equation (2.10).   □

Most computer software only supplies means of fitting log linear models and, perhaps, binary logistic models. With the former, the user is then free to choose, where applicable, the appropriate response variable(s) and the parameter estimates will be correctly calculated.

Relationship to logistic models   Recall the algorithmic procedure that I presented for the example of the previous section. In Table 2.18, we saw how the results were symmetric in the two variables. Equation (2.11) provides a formalisation of this result. From this, we can draw an interesting, and important, conclusion. For a given set of data, suppose that we fit both a logistic and a log linear model, each containing the same variables. Then, parameters referring to the same relationships among these variables will have the same estimates, except that, if the logistic model has a binary response variable, they will be twice as large.

Thus, $\widehat{\mu}_i$ and $\widehat{\alpha_{ij}}$, whether obtained from Equation (2.9) or (2.11), will be the same. If the variable indexed by i has only two categories, they will also be the same as $\widehat{\mu}$ and $\widehat{\alpha_j}$ obtained from Equation (2.4), with mean constraints, except for a factor of two. In the same way, parameters with the same symbols in Equations (2.10) and (2.12) will have the same estimates, and those from Equation (2.6) will be twice as large.

Example (continued)
Let us recalculate the parameters for the study of listeners to classical music in Table 2.11 using a log linear model in place of the binary logistic model. For Equation (2.12), we obtain the estimates, $\widehat{\mu_1} = -0.383, \widehat{\phi_1} = 0.085, \widehat{\psi_1} = -0.029,$ $\widehat{\alpha_{11}} = 0.044, \widehat{\beta_{11}} = 0.224, \widehat{\delta_{11}} = -0.255,$ and $\widehat{\gamma_{111}} = 0.166$. We see that those parameters referring to the same relationships among variables, indicated here by their having the same Greek letters, have estimates that are exactly one-half those that we previously obtained.   □

This relationship between logistic and log linear models arises from the fact that the marginal totals are not used in calculating the estimates. It means that any variable can arbitrarily be chosen as the response, and the estimates of the relationships between variables will still be correct. This is the only statistical model with this special characteristic. It is of particular importance in retrospective designs, such as case-control studies (Section 1.4.3), where things have been done 'backward'. Only the logistic and log linear models will provide correct estimates of the relationships among the variables, as we saw in the analysis of the data on œsophageal cancer in Table 2.21.

Types of independence A model that has as many equations to solve as there are unknown parameters to estimate is said to be *saturated*. The only model that I have presented so far that was not saturated was the logistic regression model of Section 2.2.5. In complex models, such as log linear models with several variables, we usually wish to simplify things by eliminating unnecessary parameters (as I did in Table 2.21). We must wait until Chapter 3 for criteria to determine what is 'unnecessary'. However, we can already look at the interpretation of some of these simplified models.

In Section 2.2.4, we already saw that there are general limits to ways in which complex models can be simplified. If some interaction is required in a model, then generally all simpler terms contained within it should also be retained. In this way, we keep the hierarchical nature of the model. I shall follow this criterion in outlining a series of possible simplifications below.

When various parameters in a log linear model are set to zero, different types of independence result. It is useful to look at the interpretations of these different possibilities. Let us look at possible simplifications of the log linear model with three variables, that in Equation (2.12).

- If a relationship exists between each pair of variables separately, without these relationships being influenced by the third variable, then $\gamma_{ijk} = 0$ for all i, j, k:

$$\log\left(\frac{\pi_{ijk}}{\dot{\pi}}\right) = \mu_i + \phi_j + \psi_k + \alpha_{ij} + \beta_{ik} + \delta_{jk} \tag{2.13}$$

- If a pair of variables, say those indexed by i and j, are independent for each value of the third variable, then $\alpha_{ij} = \gamma_{ijk} = 0$ for all i, j, k:

$$\log\left(\frac{\pi_{ijk}}{\dot{\pi}}\right) = \mu_i + \phi_j + \psi_k + \beta_{ik} + \delta_{jk} \tag{2.14}$$

This is called conditional independence, because it depends on the value of the third variable. There are three versions of this model, depending on which set of parameters is zero.

- If any one variable, say that indexed by k, is completely independent of the other two jointly, then $\beta_{ik} = \delta_{jk} = \gamma_{ijk} = 0$ for all i, j, k:

$$\log\left(\frac{\pi_{ijk}}{\dot{\pi}}\right) = \mu_i + \phi_j + \psi_k + \alpha_{ij} \tag{2.15}$$

This is equivalent to a model for the relationship between only two variables, as in Equation (2.11). There are also three versions of this model.

- If all three variables are completely independent of each other, then $\alpha_{ij} = \beta_{ik} = \delta_{jk} = \gamma_{ijk} = 0$ for all i, j, k:

$$\log\left(\frac{\pi_{ijk}}{\dot{\pi}}\right) = \mu_i + \phi_j + \psi_k \tag{2.16}$$

All these possibilities do not necessarily make sense in all situations. Thus, the parameters for margins corresponding to explanatory variables in logistic and log linear models should not be set to zero. For example, for log linear models, the minimal model is Equation (2.16) if all three variables are responses or only one is explanatory, whereas it is Equation (2.15) if those indexed by i and j are explanatory and that by k the response. None of the parameters in the appropriate one of these equations should be set to zero. In contrast, for a logistic model, such as Equation (2.6), only μ could generally not be set to zero.

The same general types of interpretation extend to models with more variables. Because of the relationship described above between the two types of models, these forms of dependencies also hold for logistic models, where appropriate.

Unfortunately, because these simplified models are not saturated, they cannot be fitted directly by solving a set of equations. For example, for three variables, the number of equations stays the same as for the model of Equation (2.12), but the number of unknown parameters to be estimated has been reduced. Thus, computer software is required to obtain the estimates.

2.3.3 LOG LINEAR REGRESSION

Quantitative explanatory variables can also be introduced into log linear models, in a way similar to logistic regression. Again, I shall only look at the simplest case. Here, it will be a one-way table of frequencies classified by a quantitative variable. If the probabilities are π_i, let us set $\nu_i = n_\bullet \pi_i$, the mean number in category i for a sample of size $n_\bullet$. Then, we can write a *log linear regression model* as

$$\log(\nu_i) = \beta_0 + \beta_1 x_i \tag{2.17}$$

where x_i is usually the centre of the interval for the value of category i of the variable. This model is also sometimes called Poisson regression (Section 4.3.1). As with logistic regression, we can solve for the mean,

$$\nu_i = e^{\beta_0 + \beta_1 x_i} \tag{2.18}$$

to give an equation that can be plotted.

Because the model involves I equations with only two unknown parameters, it is not saturated. Thus, we cannot solve the equations. A method similar to that for logistic regression can be used to estimate the parameters approximately. Only the way of obtaining z_i and w_i changes: here, we have

$$z_j = \log(n_i)$$
$$w_j = n_i$$

We again calculate

$$A = \sum_j w_j$$
$$B = \sum_j w_j z_j$$
$$C = \sum_j w_j x_j$$
$$D = \sum_j w_j x_j^2$$
$$E = \sum_j w_j x_j z_j$$

We then use the same equations as for logistic regression to obtain the parameter estimates:

$$\widehat{\beta}_1 \doteq \frac{E - B \times C/A}{D - C^2/A}$$
$$\widehat{\beta}_0 \doteq \frac{B}{A} - \widehat{\beta}_1 \frac{C}{A}$$

Example
Table 2.22 gives the counts of accidents to men working in a soap factory over a five-month period. For a log linear regression, the approximate parameter estimates are $\widehat{\beta}_0 \doteq 5.36$ and $\widehat{\beta}_1 \doteq -0.609$. The histogram of the data and this regression model are plotted together in Figure 2.8. Not surprisingly, the mean number of men decreases with increasing number of accidents; a lot of men have no accidents and few men have many accidents.

We see that the model is fairly close to the observed frequencies, although we have only estimated two parameters instead of 14, one for each category of the variable. Although $\beta_0 + \beta_1 x_i$ describes a straight line, the fitted line is curved because of the logarithm on the left-hand side of Equation (2.17). $\qquad\qquad \square$

As for logistic regression, exact estimates can be obtained by iterating with weights $w_i = \widetilde{\nu}_i$, obtained from Equation (2.18) of the previous step. This is done using computer software.

Table 2.22. Counts of accidents to men working in a soap factory over a five month period. (Irwin, 1975, from Newbold)

Number of accidents	Number of men	Number of accidents	Number of men
0	239	7	1
1	98	8	0
2	57	9	4
3	33	10	1
4	9	11	0
5	2	12	0
6	2	13	1

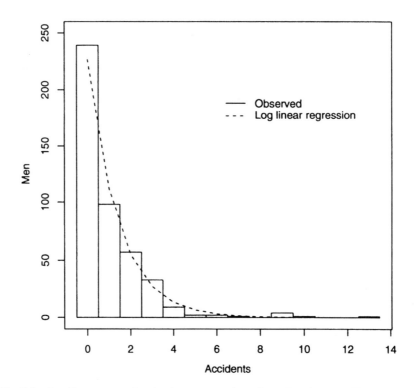

Fig. 2.8. Log linear regression for the mean number of men involved in different numbers of accidents in a soap factory.

Table 2.23. Returns from a postal survey for the number of occupants in each house. (Lindsey and Mersch, 1992)

Occupants	1	2	3	4	5	6	7
Houses	436	133	19	2	1	0	1

Example (continued)
For the accident data, the exact estimates are $\widehat{\beta_0} = 5.42$ and $\widehat{\beta_1} = -0.706$.   □

One important application of regression models is for prediction. Care, however, must be taken to verify that the predicted values are produced under suitable conditions.

Example
In a postal survey, respondents were asked the number of people residing in the house, as presented in Table 2.23. Evidently, no houses with zero occupants will be involved. (Why?) The approximate parameter estimates are $\hat{\beta_0} \doteq 7.43$ and $\hat{\beta_1} \doteq -1.34$ and the exact estimates are $\hat{\beta_0} = 7.53$ and $\hat{\beta_1} = -1.43$. The mean number of houses decreases with increasing number of occupants.

Suppose now that we want to predict the number of houses with no occupants. This is given by setting $x_i = 0$ in Equation (2.18) so that

$$\tilde{\nu}_i = e^{7.53}$$
$$= 1872.3$$

This is clearly an unreasonable value.   □

Predicting using values of the explanatory variable outside the range of those observed is called *extrapolation*. One must be sure that the model still holds under these unobserved conditions!

2.3.4 ORDINAL RESPONSE

I have now presented models fulfilling six of the seven goals set out at the beginning of Section 2.2. The final group of models in this chapter will be for data where the response variable has more than two categories and these are ordered. In other words, the variable is ordinal, as are those in Table 2.24.

Although the polytomous logistic and log linear models can be applied to such data, they do not take into account this particular structure, the ordering. Nevertheless, in such cases, the estimated parameter values should reflect the order. However, it is usually preferable to construct models that explicitly incorporate such structure.

Each ordinal model imposes an ordering on the data in some specific way. Here, I shall look at two of the most common approaches.

Continuation ratio model   First suppose that the ordinal response represents a direction through which respondents may move, one step at a time. Then, if moves are from left to right on the ordinal scale, we may be interested in the probability of a move one further step, given that an individual is already at a given position on the scale. Notice that the model will not be the same if moves are from right to left.

This happens to be the simplest ordinal response model to construct. It results in a new binary response variable comparing the first category of response with the second, then the first two categories to the third, the first three to the fourth, and so on. Thus, this new response variable always compares all categories lower in order than a certain point with the next higher category, (or those higher with the next lower, if we start from the right). This procedure forces an order on the variable. The result is called the *continuation ratio model*. It can be analysed directly as a binary logistic model, after the table is appropriately reconstructed.

Consider the case of an ordinal response variable depending on only one explanatory variable, although the procedure is directly extensible to several. If the ordinal has I categories, $I - 1$ new subtables will be formed. Each will have two columns; these will correspond to the new binary response.

The two columns of the first subtable contain the first two columns of the original table unchanged, comparing the first response category with the second. The second subtable contains the sum of the first two columns and the third column of the original table, comparing the first two categories with the third. If the ordinal response has more than three categories, the third subtable contains the sum of the first three columns and the fourth column of the original table, and so on.

Now, a binary logistic model with two explanatory variables can be applied to the new table. One explanatory variable is the original one of interest, whereas the other indexes the newly created subtables. The interaction between them should be set to zero, but this will only be possible if computer software is used. In any case, if the model is appropriate, the interaction between these two variables should be about zero.

Example

Consider the classification of long-term schizophrenic patients in two London mental institutions according to the length of stay and visiting rights. The data are given in Table 2.24 and the corresponding histograms presented in Figure 2.9. Here, both variables might be taken to be ordinal, but I shall only consider the type of visit as the ordinal response.

The data are reconstructed, as described above for a continuation ratio model, in Table 2.25. With a three-category response, there are two new subtables. The parameters of interest refer to the length of stay. They are estimated to be $\hat{\alpha} = (1.327, -0.115, -1.211)$, without setting the interaction to zero. They indicate diminishing chance of being in the lower category, of relatively more visits, as the length of stay increases. Notice how the parameter values for this ordinal explanatory variable are fairly evenly spaced, although no structure has been

Table 2.24. Length of institutionalisation of schizophrenic patients (years) and visiting rights. The types of visit are (A) goes home or visited regularly, (B) visited less than once a month and does not go home, and (C) never visited and never goes home. (Haberman, 1974, from Wing)

Length of stay	Type of visit		
	A	B	C
2 – 10	43	6	9
10 – 20	16	11	18
> 20	3	10	16

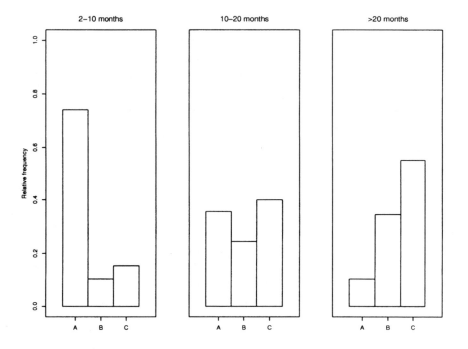

Fig. 2.9. Histograms for type of visit (A: goes home or visited regularly, B: visited less than once a month, C: never visited and never goes home) for the three lengths of stay.

imposed on them.

If computer software for fitting logistic models is available, the parameter estimates, with the interaction set to zero, are $\hat{\alpha} = (1.251, -0.181, -1.070)$, yielding the same conclusions. The interactions appear to be sufficiently close to zero for the model to be appropriate.   □

Proportional odds model   In the second model, moves can be in both directions and are not restricted in distance. Thus, the model will be symmetric in the two

Table 2.25. Table 2.24 reconstructed for the continuation ratio model.

Length	Type of visit	
of stay	A	B
2 – 10	43	6
10 – 20	16	11
> 20	3	10
	A+B	C
2 – 10	49	9
10 – 20	27	18
> 20	13	16

directions.

Fitting this model involves reconstructing the original data table in a slightly different way. Now, we compare the first category of response with all the following ones, then the first two categories with the remaining ones, the first three with the remaining, and so on. Thus, the new response variable always compares all categories lower in order than a certain point with all those higher than that point, again forcing an order on the variable. This is called the *proportional odds model*. Unfortunately, it can only be analysed approximately as a binary logistic model, after the table is appropriately reconstructed. Exact estimation requires special computer software.

Again, take the case of one explanatory variable. Here, the two columns of the first subtable contain the first column and the sum of all other columns of the original table. The second subtable contains the sum of the first two columns and the sum of the remaining columns and so on. Again, a binary logistic model with two explanatory variables can be applied to the new table, but now only approximate estimates will be obtained. The interpretation, and the results, will often be similar to those for the continuation ratio model.

Example (continued)
The data on schizophrenics, from Table 2.24, are reconstructed appropriately for the proportional odds model in Table 2.26. Here, the approximate estimates of the parameters of interest, again without setting the interaction to zero, are $\hat{\alpha} \doteq (1.342, -0.127, -1.215)$. This also indicates decreasing chance of being in the lower category, of relatively more visits, as the length of stay increases.

If special computer software for this model is available, the exact estimates are $\hat{\alpha} = (1.349, -0.265, -1.084)$. $\quad\quad\square$

Both these models for ordinal variables can also be applied when several explanatory variables are presented. Each time, the reconstructed contingency table will have one more dimension than the original table. When there is a small number of categories of the ordinal scale, as in the example, the continuation ratio and proportional odds models will usually give similar results: possible moves are re-

Table 2.26. Table 2.24 reconstructed for the proportional odds model.

Length	Type of visit	
of stay	A	B+C
2 – 10	43	15
10 – 20	16	29
> 20	3	26
	A+B	C
2 – 10	49	9
10 – 20	27	18
> 20	13	16

stricted and will be similar in the two models. With more categories, differences between them can become more apparent.

Other models are also available for ordinal variables, including ordinal explanatory variables. They all require special computer software that is not widely available.

2.4 Exercises

(1) Find a study in the literature for which results are reported in the form of a contingency table.

    (a) Describe the measures taken by the research workers to avoid Simpson's paradox occurring.

    (b) List ways in which Simpson's paradox might have made the authors' results questionable.

    (c) Invent a binary variable not used in the study and subdivide the published table in such a way that Simpson's paradox occurs.

(2) Fit a logistic model to the following data:

    (a) the data on injuries in car accidents and wearing seatbelts of Table 1.9;

    (b) the data on myocardial infarction and contraceptive use of Table 1.10;

    (c) the data on opinions about the death penalty and gun registration of Table 2.3.

In each case, interpret the parameters and discuss the meaning of the results.

(3)   (a) Fit a log linear model to the migration data of Table 1.8.

    (b) Explain how residence in 1971 is related to that in 1966.

    (c) Notice that the four areas are ordered from the north to the south of Britain. Can a better model be constructed using this information?

(4) The table on the next page (Fienberg, 1977, p. 16) gives data on the choice of piano by soloists playing for selected major American orchestras during the 1973–1974 concert season in the USA.

	Piano	
Orchestra	Steinway	Other
Boston	4	2
Chicago	13	1
Cleveland	11	2
Minnesota	2	2
New York	9	2
Philadelphia	6	0

(a) Study the relationship between the two variables.

(b) Might a log linear model be more appropriate than a logistic model?

(c) The same soloist may have appeared with different orchestras. Discuss what difficulties this may create for the models that you have used.

(5) The table below gives the frequency of coronary heart disease by age group (Hosmer and Lemeshow, 1989, p. 4). The latter was originally measured in years, but larger groupings were then created.

	Coronary heart disease	
Age	Yes	No
20–29	9	1
30–34	13	2
35–39	9	3
40–44	10	5
45–49	7	6
50–54	3	5
55–59	4	13
60–69	2	8

(a) Fit a logistic regression model to these data.

(b) Plot and interpret the results.

(c) What difficulties would you have encountered in making the plot if you had used the original raw data with the actual ages, in years, of the 100 people involved?

(6) In Section 2.2.4, we studied data on listeners to classical music radio programmes (Table 2.11). The table on the next page gives similar data on listening to religious and to discussion programmes on the radio (Lazarsfeld, 1955, pp. 115–125).

	Education			
	High	Low	High	Low
	Listen to			
	religious programmes		discussion programmes	
Age	Yes No	Yes No	Yes No	Yes No
Old	45 355	285 615	210 180	360 540
Young	55 545	115 285	240 360	100 300

The definitions of the age and education variables are the same as stated above in Section 2.2.4. The rounded values in the tables result because of the stylised nature of the data, already mentioned above in the text.

(a) Check that the joint distribution of age and education is the same in all three tables.

(b) Fit an appropriate logistic model to each half of the table (each type of programme).

(c) Are the results similar to those given above for classical music?

(d) Can you explain why?

(7) Consider data on the relationship among delinquency, socioeconomic status (SES), and being a boy scout, given below (Agresti, 1990, p. 157).

| | | Delinquent | |
SES	Scout	Yes	No
Low	Yes	10	40
Low	No	40	160
Medium	Yes	18	132
Medium	No	18	132
High	Yes	8	192
High	No	2	48

(a) Fit a logistic model to explore the relationships among the variables.

(b) What is peculiar about these data? Look at models for the marginal tables, when delinquency depends on only one of the explanatory variables.

(c) Relate your conclusions to the difficulties in drawing causal conclusions from sample survey data.

(8) In Section 2.2.4, we looked at a study of knowledge of cancer, given in Tables 2.10 and 2.13 above. The table on the following page reproduces the data, but with the 'lectures' variable replaced by 'serious reading' (Lombard and Doering, 1947).

Newspaper	Reading	Radio			
		Yes		No	
		Knowledge			
		Good	Poor	Good	Poor
Yes	Yes	125	75	228	195
	No	43	63	82	162
No	Yes	17	19	70	91
	No	17	53	86	403

(a) Fit a logistic model and interpret the results.

(b) Compare them with those given above and explain any differences.

(c) If you have appropriate computer software available, look at models for all four explanatory variables simultaneously.

Radio	Newspaper	Lectures	Reading			
			Yes		No	
			Knowledge			
			Good	Poor	Good	Poor
Yes	Yes	Yes	23	8	8	4
Yes	Yes	No	102	67	35	59
Yes	No	Yes	1	3	4	3
Yes	No	No	16	16	13	50
No	Yes	Yes	27	18	7	6
No	Yes	No	201	177	75	156
No	No	Yes	3	8	2	10
No	No	No	67	83	84	393

(d) Again, compare the results with those from the simpler tables and explain any differences.

(9) A study was conducted to determine factors that might influence shopping behaviour. The sample was taken at random from the population of the town of Dukinfield, Greater Manchester, England. In the following table, the variables choice of shopping centre, age, income, and car ownership are presented (Fingleton, 1984, p. 25). Unfortunately, the author does not state how the categories for the variables were constructed.

Age	Income	Car owner			
		Yes		No	
		Shopping centre			
		Near	Other	Near	Other
Young	Low	12	57	17	48
	High	3	24	2	3
Old	Low	18	53	51	105
	High	2	11	1	0

(a) Fit a logistic model.

(b) Interpret the results.

(10) The following table shows the numbers of household burglaries in Detroit, USA, 1974–1975, obtained from the National Crime Survey (Nelson,
1980).

Number of burglaries	Number of households
0	8385
1	976
2	183
3	35
4	5
5	2

(a) Fit a log linear regression model to these data.

(b) Plot and interpret the results.

(11) The table in Exercise (1.6) gave the frequency of recall of a stressful event
over an 18 month period.

(a) Study how recall depends on time by fitting a log linear regression
model.

(b) Plot and interpret the results.

(12) People involved in a driver education study were followed over a four-year
period. Traffic violations each year among male subjects in the control
group were recorded as shown in the following table (Davis, 2002, p. 228).

Year				
1	2	3	4	
No	No	No	No	731
No	No	No	Yes	310
No	No	Yes	No	256
No	No	Yes	Yes	196
No	Yes	No	No	156
No	Yes	No	Yes	121
No	Yes	Yes	No	114
No	Yes	Yes	Yes	152
Yes	No	No	No	61
Yes	No	No	Yes	40
Yes	No	Yes	No	45
Yes	No	Yes	Yes	39
Yes	Yes	No	No	47
Yes	Yes	No	Yes	42
Yes	Yes	Yes	No	46
Yes	Yes	Yes	Yes	53

(a) Develop a log linear model to describe the association between violations in the various years.

(b) Is the association stronger for years closer together in time?

(c) Is it reasonable to simplify the model by only including associations between adjacent years?

(13) The following table gives the results of a social survey of income and job satisfaction in the USA (Agresti, 1990, p. 21, from Norusis). They are taken from the 1984 General Social Survey of the National Data Program.

	Satisfaction			
Income	Very dissatisfied	Little dissatisfied	Moderately satisfied	Very satisfied
< $6000	20	24	80	82
$6000–14999	22	38	104	125
$15000–24999	13	28	81	113
≥ $25000	7	18	54	92

(a) Fit a polytomous logistic model to these data and interpret the results.

(b) Now fit your choice of ordinal model.

(c) Do you find any differences in the interpretation of the results?

(d) What are the advantages and disadvantages of fitting the ordinal model as compared with the polytomous logistic model?

(14) The table below shows party affiliation and political ideology of a sample of voters during the 1976 Wisconsin, USA, presidential primary election (Agresti, 1984, p. 87, from Hedlund).

	Political ideology		
Party	Liberal	Moderate	Conservative
Democrat	143	156	100
Independent	119	210	141
Republican	15	72	127

(a) Fit the continuation ratio and proportional odds models to these data.

(b) Compare and interpret the results.

(15) The following table gives ratings of the performance of radio and television, by the person's colour for samples taken in two different years (Agresti, 1984, p. 103, from Duncan and McRae).

		Rating		
Year	Colour	Poor	Fair	Good
1959	White	54	253	325
	Black	4	23	81
1971	White	158	636	600
	Black	24	144	224

(a) Fit the continuation ratio and proportional odds models to these data.

(b) Compare and interpret the results.

3
Inference

We have now explored some ways of constructing models to help to distinguish between random and systematic variability in an observed sample of data. We have seen how to *estimate* certain unknown parameters by calculating values from such a sample of data. We also know that such calculated values will generally not be identical to those that we would obtain if we could calculate them for the complete population in which we are interested. It is now time to address several questions concerning how the results that we obtain for our models from a sample can be related to the population as a whole. This is called *inference* from the observed data to the population.

3.1 Goals of inference

In trying to generalise from a sample of data to some wider context, we should be clear as to the goals that we wish to pursue. Several distinct contexts can be distinguished:

- Scientific procedures search to *discover* something new. The actual process of scientific discovery is extremely difficult to formalise; it involves the development of new models or the application of old models in new contexts. One important way in which statistics can contribute to this process is by providing procedures for comparing such models, given some appropriate observed data.
- Technology is concerned with *decisions* among which available scientific discoveries may be useful, and how practically to implement them. Statistics contributes to the technological decision process by providing means of testing hypotheses. Many of these procedures were originally developed with quality control in mind.
- People often want to obtain information to help them in making *personal* decisions, for example, about investments on the stock market. The incorporation of prior information about the alternatives available has found particular importance in modern statistics in the theory of personal decision-making.

Thus, statistical inference has very wide applications. However, in the modelling approach emphasised here, I shall give the central place to scientific applications.

3.1.1 DISCOVERY AND DECISIONS

Misconceptions abound about the process of scientific discovery and the place of statistics in this process. Thus, a myth prevails in statistics, and in philosophy, that science advances by testing hypotheses. Certainly, *any* scientific theory, and its statistical representation by a model, must be specified in such a way that it can be checked ('tested') by confrontation with empirical data. However, every scientific theory, and every statistical model, is an approximation to reality. It will always be wrong in many aspects. When a model is checked by obtaining a sample of data, it will *always* be deficient in certain ways. But, unless a clearly better model is available, it will *never* be rejected outright. Thus, science advances, not by hypothesis testing, but by comparing theoretical models in the light of empirical data. In statistics, this is called *model selection*.

Hypothesis testing does, however, have an important place in statistics, though rarely in the context of scientific applications. Its role is to help in decision-making.

Example
Consider the development of a new treatment for cancer, say some drug. First, a potential molecule must be isolated. Then, preliminary study in the early phases of development involves checking toxicity of the substance and verifying that it does have the desired effect (activity). A pharmacokinetic model is usually constructed to describe how it moves through the body. These are scientific goals to discovery new information. Once this has been obtained, final phase clinical trials are conducted on a larger scale. These are designed, not primarily to learn something new, but to demonstrate to the public and to the authorities that the treatment actually does provide improvement over the traditional approach (efficacy) so that it can be marketed. This is not a scientific goal, but rather a decision problem: can the new treatment be allowed on the market or not?   □

Unfortunately, much of classical statistics employs testing tools in attempts to answer scientific model selection questions. They are not appropriate for several reasons:

(1) A decision problem must always be framed in terms of a small number of known outcomes among which a choice must be made, whereas science is a quest for the yet unknown outcome.

(2) Decisions generally imply one-time actions, whereas science is an ever-continuing interaction between models and data: models ('hypotheses') are modified and improved, or replaced, not simply rejected outright.

(3) Decisions imply calculable gains and losses that must be weighed in making the choice, whereas science does not and, by definition, cannot: gains and losses only become relevant in technological applications of scientific discoveries.

Statistics provides a wide variety of tools. Many users consider them to be general purpose tools, valid in all circumstances. This is not the case. Care must be taken to select the appropriate tool for each problem. This applies to inference procedures as well as to models.

3.1.2 TYPES OF MODEL SELECTION

Model comparison has several levels of complexity that do not necessarily correspond to the order in which they must be addressed in a scientific investigation.

In the simplest situation, we have one given model function that we assume to be adequate for the phenomenon under study. Then, model selection involves only the comparison of the parameter values in that function.

Example
Consider the logistic regression function in Equation (2.7) of Section 2.2.5:

$$\log\left(\frac{\pi_{1|j}}{\pi_{2|j}}\right) = \beta_0 + \beta_1 x_{1j}$$

We wish to judge the relative suitability of the (infinite) set of different models specified by different possible values of the parameters, β_0 and β_1. We shall want to select which of these models, that is, which of these values, are most likely given the data in our sample. □

However, often in scientific investigations, the situation is more complex. Several different model functions will usually be under consideration ('in competition').

Example (continued)
We may want to compare the above model function with

$$\log\left(\frac{\pi_{1|j}}{\pi_{2|j}}\right) = \beta_0 + \beta_2 x_{2j}$$

Here, the model function has changed, not just the parameter values in it, because a different explanatory variable is involved. The parameters in the two models are not even comparable, although some have the same names, because they have different interpretations in the two cases. We want to know which of the two functional forms is best supported by the observed data. □

When faced with a number of models, we can distinguish several questions, all conditional on the data observed:

(1) Among the model functions currently under consideration by the scientist, which are most appropriate (*model selection*)?
(2) For a given model function, what are the 'best' estimates of the unknown parameter values contained in it (*point estimation*)?

(3) Given that the 'best' parameter estimates from the data will not be identical to those in the population, what range of values of the parameters is reasonably plausible (*interval estimation*)?

(4) Does a given model adequately represent the phenomenon under study (*goodness-of-fit*)?

(5) If none of the models is satisfactory, what hints can they provide about a more appropriate one (*diagnostics*)?

These questions have a logical order in any scientific investigation. In this chapter, I shall explore all of them. However, for technical and pedagogical reasons, I shall not always be able to follow this order strictly.

Both comparing parameter values (interval estimation) in a given function and comparing functions (traditionally called model selection) are fundamentally similar. There will always be a set of models that is reasonable for the observed sample, not just one 'best' model. I shall begin by examining the first problem and showing how various models are more or less *likely* to be able to explain the phenomenon under study, given the observed data. Generally, one of these models will be most likely (the point estimate); we shall see that the estimates that we calculated in Chapter 2 have this property. This procedure will, then, provide a basis for attacking the second problem which has the added complication that some models to be compared may be more complex than others.

The fundamental statistical model selection techniques will also provide a basis for hypothesis testing (Section 3.4.3) and incorporating prior information (Section 3.4.4).

3.2   Likelihood

In the context of scientific modelling, the basic questions revolve around what we mean by one model being more plausible or likely to describe or explain the phenomenon under study than another, given the observed data. We shall require a precise definition of this concept. To accomplish this, I shall start with the simple situation where we have only one model function and wish to determine which parameter values are plausible.

When a sample is chosen from the population, one that does not contain every individual in that population, the estimates of any parameters calculated for the sample will not generally be the same as the unknown values for the entire population. Even if you flip an unbiased coin ten times, you would not always expect to obtain five heads! No heads at all will even occasionally turn up. The same kind of variation in estimates of parameters calculated from samples will occur in more complex situations. In this first simple case, what we want to know is, given the model function and the observations made, what is the range of plausible values of the parameter for the population?

3.2.1   LIKELIHOOD FUNCTION

Let us consider the case where we have some simple model function for our population, containing one unknown parameter.

Probabilities before collecting data   If we fix the unknown parameter at some arbitrary value, we can use the function to calculate the probabilities of the various possible observations that could be made. For different values of the parameter, the probability of observing some specified value of the response will be different.

Example
Suppose that we are planning a study involving answers to some question, with possible responses yes or no. This will yield a binary variable. Let us assume the simple model whereby it follows a binomial distribution. Then, the unknown parameter is the probability π_1 of, say, a yes answer. From Section 1.3.4, we know that the probability calculated for a sample of independent observations is the product of the probabilities of the individual responses.

   If we fix this parameter at some arbitrary value, we can calculate the probabilities of different numbers n_1 of yeses in a sample of fixed size using the results from Section 1.3.4. If $\pi_1 = 0.2$, the probability of observing 6 yeses in 10 questionnaires will be different than if $\pi_1 = 0.5$. Both can be calculated using Equation (1.9):

$$\Pr(n_1; \pi_1, n_\bullet) = \binom{n_\bullet}{n_1} \pi_1^{n_1} (1 - \pi_1)^{n_\bullet - n_1}$$

In the function $\Pr(n_1; \pi_1, n_\bullet)$, any symbols following the semicolon are taken to have fixed values; only those before the semicolon are allowed to vary.   □

   Thus, we can fix the values of all parameters in any model and use the model to calculate the probabilities of observing different possible sets of observations. This procedure is important as a theoretical step before the observations are made.

Example (continued)
For a sample of $n_\bullet = 10$ from a binomial distribution, we can construct a table of possible results for various different models. Those for a few models are given in Table 3.1. For a given model, that is, a given value of π_1, we read down a column to see the probabilities of the 11 possible sample results. As might be expected, the larger is the probability parameter π_1, the more probable is a sample with many yeses. Notice also that the highest probability, which is underlined in the table, is for a sample with the same proportion of yeses as the theoretical probability π_1. However, other samples nearby have almost as high a probability, and will be observed almost as often.   □

Table 3.1. Binomial probabilities for a sample of ten observations under different models.

n_1	π_1								
	0.1	0.2	0.3	0.4	0.5	0.6	0.7	0.8	0.9
0	0.349	0.107	0.028	0.006	0.001	0.001			
1	0.387	0.268	0.121	0.040	0.010	0.002			
2	0.194	0.302	0.233	0.121	0.044	0.011	0.001		
3	0.057	0.201	0.267	0.215	0.117	0.042	0.009	0.001	
4	0.011	0.088	0.200	0.251	0.205	0.112	0.037	0.006	
5	0.002	0.026	0.103	0.201	0.246	0.201	0.103	0.026	0.002
6		0.006	0.037	0.112	0.205	0.251	0.200	0.088	0.011
7		0.001	0.009	0.042	0.117	0.215	0.267	0.201	0.057
8			0.001	0.011	0.044	0.121	0.233	0.302	0.194
9				0.002	0.010	0.040	0.121	0.268	0.387
10				0.001	0.001	0.006	0.028	0.107	0.349

Likelihoods after collecting data   Now, once we make the observations, we are in a different situation. We have observed certain values of the variable(s) and are no longer interested in predicting what might be observed. We *know* what has been observed. Thus, we now place ourselves in the radically different position of saying that we do not know what the parameter values in the model are for the population, but want to determine what our observations can tell us about them. We are still making a number of theoretical assumptions, for example, that the distribution is binomial and that we only want to learn about, that is, to make inferences about, the value of π_1.

In summary, before making our observations, we take the values of the variable to be unknown and we can set the values of the parameters for theoretical reasons. After the observations are available, we know the values of the variable observed, and we take the values of the parameters to be unknown. We thus now have a function of the parameters given fixed values of the data instead of a function of the data given fixed values of the parameters. This is called the *likelihood function*. For independent observations, it will be a product of factors for each individual response.

Example (continued)
For the binomial distribution, we can write this relationship as

$$\Pr(n_1;\pi_1,n_{\bullet}) = L(\pi_1;n_1,n_{\bullet})$$

where we use the left-hand side before collecting the data and the right-hand side after obtaining the data. The former is a probability function and the latter is the corresponding likelihood function. In both cases, the semicolon separates fixed values on its right from values that can change on its left.   □

Once we have the data in hand, we shall be interested in comparing different models that might describe how they were generated. For each such model, we can calculate the probability of our observations. In other words, we can vary the value of the parameter, that is, consider different models, and see how probable our given observations are for each. Then, we can conclude that *a value of the parameter that would make our observations more probable is more plausible or more likely, according to those data.*

Example (continued)

If we observed 6 yeses in a sample of 10 questionnaires, we can compare, among others, the models with $\pi_1 = 0.5$ and $\pi_1 = 0.2$, assuming a binomial distribution:

$$L(\pi_1 = 0.5; n_1 = 6, n_\bullet = 10) = \Pr(n_1 = 6; \pi_1 = 0.5, n_\bullet = 10)$$
$$= \binom{10}{6} 0.5^6 0.5^4$$
$$= 0.205$$
$$L(\pi_1 = 0.2; n_1 = 6, n_\bullet = 10) = \Pr(n_1 = 6; \pi_1 = 0.2, n_\bullet = 10)$$
$$= \binom{10}{6} 0.2^6 0.8^4$$
$$= 0.006$$

Both values can also be obtained directly from Table 3.1. Thus, the model with $\pi_1 = 0.5$ makes the observations $0.205/0.006 = 34$ times more probable than that with $\pi_1 = 0.2$. We can conclude that the first model is 34 times more plausible or likely than the second, given the observed data.

If we now look at Table 3.1 again, but in a different light, we shall be able to see the likelihood function. Once we have a fixed, observed sample, we must look across a row, not down a column. Thus, for the values above, the row for $n_1 = 6$ gives some values of the likelihood function, with maximum at 0.6. We see that, in this table, the columns represent probability distributions and the rows likelihood functions. However, all possible values of n_1 are given, whereas only a few of the possible values of π_1 are shown.   □

A likelihood for a given parameter value has no meaning by itself, but only as compared with those for other parameter values. This is usually shown as a ratio of two probabilities, as for the relative risk and odds. But, here the meaning is very different because different models are being compared, not different observations.

3.2.2   MAXIMUM LIKELIHOOD ESTIMATE

One value of the parameter will always make the observations most probable (at least in the models that I shall consider in this text). This value is called the *maximum likelihood estimate* of the parameter. As we shall see, it is the value that I have denoted in previous chapters by a 'hat' on a parameter. This does not mean that it is the 'true' value of the parameter for the population. Other values of the parameter will be almost as likely. However, it is the 'best' single or *point estimate* for that sample of observations.

Example (continued)
For our example binomial data, if we tried various values of the parameter in the likelihood function, we would discover that the maximum is located at $\widehat{\pi}_1 = 6/10 = 0.6$. We can also see this from Table 3.1. □

For those familiar with calculus, the maximum likelihood estimate can be obtained more generally by taking the first derivative of the likelihood function, setting it to zero, and solving. However, the procedure is easier if the logarithm of the likelihood function is used. Then, the first derivative set to zero is called the *score equation*.

Example (continued)
For the binomial distribution, the likelihood function is

$$L(\pi_1) = \binom{n_\bullet}{n_1} \pi_1^{n_1} (1-\pi_1)^{n_2}$$

For simplicity of notation, I have dropped the semicolon with the indication of dependence on the given data. Notice that this likelihood is the same as the model in Equation (1.9). Then, the log likelihood is

$$\log[L(\pi_1)] = \log\left[\binom{n_\bullet}{n_1}\right] + n_1\log(\pi_1) + n_2\log(1-\pi_1)$$

Setting the first derivative to zero, we have the score equation

$$\frac{d\log[L(\widehat{\pi}_1)]}{d\widehat{\pi}_1} = \frac{n_1}{\widehat{\pi}_1} - \frac{n_2}{1-\widehat{\pi}_1} \tag{3.1}$$
$$= 0$$

Solving, we obtain

$$\widehat{\pi}_1 = \frac{n_1}{n_1 + n_2}$$

This is just the estimate that we have previously been using. □

This result holds much more generally. Thus, the parameters calculated from a sample in Chapter 2 and indicated by a hat are indeed maximum likelihood estimates.

3.2.3   NORMED LIKELIHOOD AND DEVIANCE

The likelihood function provides only a *relative* measure of plausibility of models. It can only be interpreted by comparing the likelihood values among different models. For this reason, a point of comparison will be useful for standardisation. Thus, it is often convenient to compare the likelihoods of all possible values of a parameter with that of the maximum likelihood estimate. This can be done by using the ratio

$$R(\pi_1) = \frac{L(\pi_1)}{L(\widehat{\pi}_1)}$$

This is called the *relative* or *normed likelihood function*, because it relates each value of the parameter to the maximum likelihood estimate and because it can only take values between zero and one. (Why?)

Example (continued)
For the binomial distribution, we have

$$R(\pi_1) = \frac{\binom{n_\bullet}{n_1}\pi_1^{n_1}(1-\pi_1)^{n_2}}{\binom{n_\bullet}{n_1}\widehat{\pi}_1^{n_1}(1-\widehat{\pi}_1)^{n_2}}$$

$$= \left(\frac{\pi_1}{\widehat{\pi}_1}\right)^{n_1}\left(\frac{1-\pi_1}{1-\widehat{\pi}_1}\right)^{n_2} \tag{3.2}$$

$\square$

Notice how that complicated constant factor, which does not contain the unknown parameter, cancels out. This will always happen in the comparison of any likelihoods as ratios for two different parameter values in the same probability function. Thus, these factors are not necessary in the likelihood function, at least when comparing parameter values in the same model function.

Once again, it is often easier to work with logarithms. However, remember that $0 \leq R \leq 1$, and that logarithms of such values will all be negative. Hence, it will be convenient to change the sign of all values to make them positive, yielding the negative log normed likelihood. (I shall often just call it the negative log likelihood.) Because it is defined in this way, it only takes positive values. For independent observations, it will be a sum of terms for the individual responses because the likelihood is a product of the corresponding factors.

The negative log normed likelihood involves a difference between two log likelihoods. The bigger this difference is, the farther the model with chosen fixed value of the parameter is from that with the most likely value, according to the data, and the more unacceptable is that value. The negative log likelihood provides a measure of distance between some model of interest and the most likely model, given the observed data.

For traditional reasons (see Section 3.4.3), the negative log normed likelihood is often multiplied by two. For a fixed parameter value, this is called the *deviance*:

$$D(\pi_1) = -2\log[R(\pi_1)]$$
$$= -2\{\log[L(\pi_1)] - \log[L(\widehat{\pi}_1)]\} \tag{3.3}$$

For independent observations, this will also be a sum of terms. Both the negative

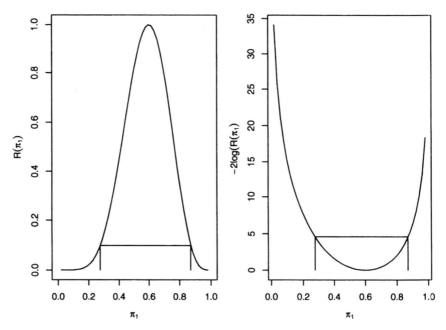

Fig. 3.1. Normed likelihood and deviance graphs for six yeses in ten binomial observations, with the 10% likelihood interval.

log likelihood and the deviance are used in the literature so that care should be taken to verify which is being used in a given context.

Likelihood intervals   When a model contains only one parameter, the preferable way to determine intervals of plausible values for it is to plot the normed likelihood function. Other possibilities are to plot its logarithm or the deviance. The width of such graphs shows the *precision* of the parameter estimate; a narrower graph indicates a more precise estimate.

Example (continued)
For our example binomial data, the normed likelihood function and the deviance are plotted in Figure 3.1. These correspond to the values for the row $n_1 = 6$ in Table 3.1, but with many more points plotted. Here, we can easily read off the maximum likelihood estimate as 0.6 and see the range of reasonable values around it.   □

As can be seen from Figure 3.1, the range of likely values is easy to visualise at a glance using the normed likelihood than the deviance because the former stops decreasing at zero whereas the latter keeps increasing towards infinity on both sides.

As one summary of precision, we may choose to say that all values of the parameter are reasonably plausible if they make the observations at least some selected fraction as probable as the most likely model for the data, the maximum likelihood estimate, does. In other words, we may take as being likely all values of the parameter that make R greater than that fraction. Naturally, the smaller the value of R that we choose, the wider will be the interval.

This value of R will correspond to a deviance of twice the logarithm of that fraction. We might also start from some fixed deviance value. The corresponding normed likelihood will then be given by $\exp(-D/2)$, where D is the deviance.

Example (continued)
Suppose that we choose, as our criterion, a 10% likelihood interval. This corresponds to deviances less than 4.6. We see from the graphs in Figure 3.1 that the interval of likely values is about $(0.275, 0.870)$. Neither the curve nor the interval is symmetric around the maximum value. More values are likely that are smaller than that point estimate than larger ones.   □

Such likelihood intervals, or approximations to them, are often called *confidence* (Section 3.4.3) or *credibility* (Section 3.4.4) *intervals*. They provide a summary indication of the precision with which a parameter is estimated.

Care must be taken not to be misled into believing that, because one parameter estimate is closer than another to a particular value of interest, it is more likely to have that value.

Example
A parameter that is closer to zero may be less likely to be zero than one farther away, as illustrated in Figure 3.2. There, the parameter with maximum likelihood estimate of 0.26 could perhaps be zero, whereas that with an estimate of 0.16 is less likely to be zero.   □

The precision of the estimate, as indicated by the width of the likelihood curve, is very important. As might be expected, the most important factor determining this precision is the number of observations made, that is, the sample size (Section 3.6). Generally, the more observations, the more information we have about the unknown parameter and the narrower will be the likelihood curve.

Example (continued)
For our binomial example, suppose that we repeated the study with 100 questionnaires and obtained 61 yeses. (We would not expect to obtain exactly 60. Why?) The normed binomial likelihood function, along with our previous one, are plotted in Figure 3.3. We see how the curve becomes narrower when more observations are available. On the other hand, if 20 yeses out of 100 had been obtained, the curve would still be about as narrow, but would be displaced with maximum at 0.2.   □

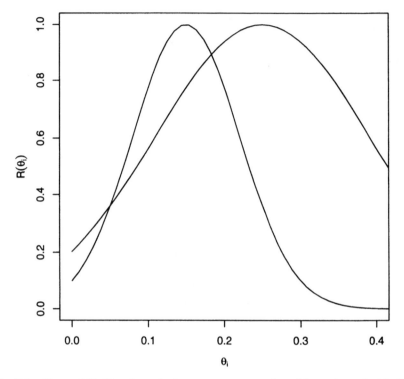

Fig. 3.2. Normed likelihood graphs for two parameters, θ_1 and θ_2, estimated with different precisions.

Things are more complicated when there are several parameters, as for logistic and log linear models for categorical data. If there are only two, the likelihood surface can be plotted, either in three dimensions or as contours, although this usually requires sophisticated statistical software. Another solution is to fix one parameter of interest at a series of values and, for each, calculate the maximum likelihood estimates of all the others. This *profile likelihood function* for one parameter can then be plotted.

Example (continued)
In Section 2.2.5 (Table 2.14), we looked at the relationship between infant malformations and their mother's alcohol consumption. We found the slope to be estimated as $\widehat{\beta}_1 = -0.317$. The exact profile likelihood curve for this parameter can be obtained with a computer by setting a series of arbitrary values of the slope and refitting the model for each. The result is plotted as the solid curve in Figure 3.4. The solid and dotted vertical lines show two possible plausibility intervals: $(-0.47, -0.11)$ and $(-0.52, -0.01)$. □

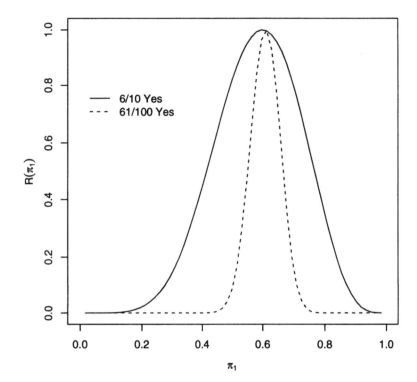

Fig. 3.3. Normed likelihood graphs for 6 yeses in 10 and 61 in 100 binomial observations.

Plausibility intervals are only summaries of the information in the likelihood function. It is always preferable, if possible, to report the likelihood function itself, or at least to plot profile likelihood curves for important parameters.

3.2.4 STANDARD ERRORS

In Section 1.2.2, we saw that a symmetric measure of variability of any quantity calculated from the data is the standard error, the square root of the variance of that quantity. Maximum likelihood estimates are such quantities and hence have standard errors. Standard errors are usually produced by statistical software when maximum likelihood estimates are calculated.

Example (continued)
Let us derive the standard error of $\hat{\pi}_1 = n_1/n_\bullet$ from a binomial distribution. In Section 1.3.4, we saw that the variance of n_1 is $n_\bullet \pi_1 (1 - \pi_1)$. From the rules for calculating variances in Section 1.2.2, the variance of $\hat{\pi}_1$ is then

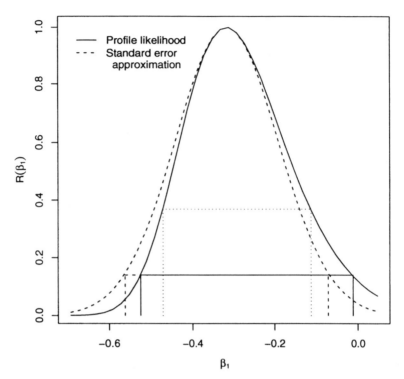

Fig. 3.4. Normed profile likelihood graph, and its approximation using the standard error, for the slope of the logistic regression for the malformations of Table 2.14. The corresponding vertical lines delimit the 95% confidence interval and the dotted lines give an interval corresponding to the AIC.

$$\frac{n_\bullet \pi_1 (1 - \pi_1)}{n_\bullet^2} = \frac{\pi_1 (1 - \pi_1)}{n_\bullet}$$

Thus, the required standard error is $\sqrt{\pi_1 (1 - \pi_1)/n_\bullet}$. However, because we do not know the value of π_1, to use it, we must substitute in $\widehat{\pi}_1$. □

For those who know calculus, the general method to calculate a standard error is first to obtain the second derivative of the negative log likelihood. This value is called the *Fisher information*. Then, the standard error is the square root of its reciprocal. The larger is the Fisher information, the smaller the variance, and the standard error, hence the more information is available about the parameter.

Example (continued)
We can now verify that this procedure yields the same answer as above for the standard error of $\widehat{\pi}_1 = n_1/n_\bullet$ from a binomial distribution. From Equation (3.1)

for the score equation of the binomial distribution, we have

$$\frac{d^2\{-\log[L(\pi_1)]\}}{d\pi_1^2} = \frac{n_1}{\pi_1^2} + \frac{n_2}{(1-\pi_1)^2}$$

We saw in Section 1.3.4 that the expected value of n_1 is $n_\bullet\pi_1$. If we substitute this into the above expression, we obtain

$$\frac{n_\bullet}{\pi_1} + \frac{n_\bullet}{1-\pi_1} = \frac{n_\bullet}{\pi_1(1-\pi_1)}$$

This is the expected Fisher information for $\widehat{\pi}_1$. Then, the square root of the reciprocal of this is its standard error, as above. □

Approximating the likelihood function Let us call z the distance of the value of some parameter, say θ, from its maximum likelihood estimate $\hat{\theta}$, measured in terms of the standard error, $\text{se}(\hat{\theta})$:

$$z = \frac{\theta - \hat{\theta}}{\text{se}(\hat{\theta})}$$

Then, an approximation to the normed likelihood function is

$$R(\theta) \doteq e^{-\frac{z^2}{2}} \tag{3.4}$$

from which approximate likelihood intervals can be obtained. Equivalently, an approximation to the deviance will be

$$z^2 = \frac{(\theta - \hat{\theta})^2}{\text{var}(\hat{\theta})}$$

where $\text{var}(\hat{\theta})$ is the variance of the maximum likelihood estimate.

Example
An approximate 13.5% likelihood interval, or deviance of 4.0, will be given by the maximum likelihood estimate plus and minus two times the standard error. □

Several values for this approximation are given in Table 3.2.

Examples (continued)
For our example binomial data with $n_1 = 6$, the maximum likelihood estimate is $\widehat{\pi}_1 = 0.6$. The estimated standard error is $\sqrt{0.6 \times 0.4/10} = 0.155$. Thus, the approximate 13.5% interval is $(0.29, 0.91)$. This compares with $(0.29, 0.85)$ for the exact interval, which can be obtained from Figure 3.1 above.

Table 3.2. Standard errors with approximate likelihood intervals and deviances.

Standard errors	Likelihood interval	Deviance
1.0	0.607	1.00
1.5	0.325	2.25
2.0	0.135	4.00
2.5	0.044	6.25
3.0	0.011	9.00
3.5	0.002	12.25

For the slope parameter of the malformation data in Table 2.14, the approximate likelihood is plotted as the dashed curve in Figure 3.4. Notice that it is necessarily symmetric around the maximum likelihood estimate, whereas the exact profile likelihood was not.   □

A standard error can be used as a quick guideline. However, because this procedure always yields a symmetric interval about the maximum likelihood estimate, it can be a bad approximation if the exact likelihood interval is asymmetric. The approximation also depends on the number of observations being sufficiently large, so that great care should be taken in using it.

3.3 Two special models

The likelihood function only provides the means to make *relative* judgements about models. Hence, whenever possible, it is important to have some points of comparison. In any modelling situation, two models will be of special interest: the most complicated and the simplest models possible or envisaged.

3.3.1 SATURATED MODELS

For any set of data, there will be at least one model that will fit the data perfectly. It will generally have as many parameters as observations and thus will often be too complicated to be of much use. As we saw in Section 2.3.2, this is called a *saturated model*.

Nevertheless, in Chapter 2, all the logistic and log linear models for which we calculated parameters, except the regression (Sections 2.2.5 and 2.3.3) and ordinal (Section 2.3.4) models, were saturated. They had as many parameters, and equations, as observations, which allowed us to solve them easily. We did, however, see in Section 2.3.2 ways in which log linear models could be simplified. Thus, for example, a model for independence is not saturated.

A saturated model can often provide an important point of comparison when we try to simplify. We can ask if our simplified model is still close enough to the data as compared with the most closely fitting one. I shall discuss this further in Section 3.5.1 below.

3.3.2 NULL MODELS

Often, one set of parameter values is of particular interest. We wish to determine if these parameters are essential to the model, given the data, or if they can be eliminated.

Example
In a logistic model, such as Equation (2.4), we shall usually be interested in whether or not the response actually depends on the explanatory variable. If not, we have independence, in which case $\alpha_j = 0$ for all j.   □

If there are several explanatory variables, several different sets of parameters may be fixed at zero, as we saw in Section 2.3.2. The simplest such model, when all possible parameters are set to zero, is called the *null model*. Examining the normed likelihood for such a model will indicate if it is reasonable, in light of the data, as compared with the best model. This will also give us some idea of the shape of the likelihood function, that is, of the precision of the estimate, because we shall now have two points on it: those for the maximum likelihood estimate and for the zero value.

Example
In a two-way $I \times J$ table, modelled by Equation (2.9), we shall be interested in the model for independence. This has $\alpha_{ij} = 0$ for all i, j.

The likelihood function for the saturated model can be written as a product of independent conditional multinomial distributions, one for each category of the explanatory variable:

$$L(\pi_{i|j}, \text{all } i, j) = \prod_j \binom{n_{\bullet j}}{n_{1j} \cdots n_{Ij}} \prod_i \pi_{i|j}^{n_{ij}}$$

Then the normed likelihood function is

$$R(\pi_{i|j}, \text{all } i, j) = \prod_j \prod_i \left(\frac{\pi_{i|j}}{\widehat{\pi_{i|j}}} \right)^{n_{ij}}$$

Under independence, the set of (conditional) probabilities will be the same for all j, that is, for all categories of the explanatory variable: $\pi_{i|j} = \pi_{i\bullet}$. The respective maximum likelihood estimates for the two models are

$$\widehat{\pi_{i|j}} = \frac{n_{ij}}{n_{\bullet j}}$$

$$\widehat{\pi_{i\bullet}} = \frac{n_{i\bullet}}{n_{\bullet\bullet}}$$

Substituting these into the normed likelihood gives

$$R(\alpha_{ij} = 0, \text{all } i, j) = R(\pi_{ij} = \widehat{\pi_{i\bullet}}, \text{all } i, j)$$

$$= \prod_j \prod_i \left(\frac{\widehat{\pi_{i\bullet}}}{\overline{\pi_{i|j}}} \right)^{n_{ij}} \qquad (3.5)$$

$$= \prod_j \prod_i \left(\frac{n_{i\bullet} n_{\bullet j}}{n_{\bullet\bullet} n_{ij}} \right)^{n_{ij}}$$

Then, the negative log normed likelihood is

$$-\log[R(\alpha_{ij} = 0, \text{all } i, j)] = -\log[R(\pi_{ij} = \widehat{\pi_{i\bullet}}, \text{all } i, j)]$$

$$= \sum_i \sum_j n_{ij} \log(n_{ij}) + n_{\bullet\bullet} \log(n_{\bullet\bullet})$$

$$- \sum_i n_{i\bullet} \log(n_{i\bullet}) - \sum_j n_{\bullet j} \log(n_{\bullet j}) \qquad (3.6)$$

and the deviance is twice this. Notice that, in contrast to the (log) odds, this can be calculated even when some frequencies are zero, because $0 \log(0) = 0$.   □

In some books, a value of the deviance calculated in this way for a contingency table is called G^2.

Example (continued)
Let us apply these results to look at independence for the data on graduate plans in Table 1.3. We obtain a normed likelihood of $R(\alpha_1 = 0) = 0.41$ whose negative logarithm is 0.88. This indicates that a model of independence is reasonable for these data. The best model for these data makes them only about 2.5 times as probable as the independence model. If this were a sample, the differences observed between the sexes could just be due to chance in the random selection of the sample from a data generating mechanism that was identical for both sexes.□

In this development of the normed likelihood for a two-way table, I fixed $\alpha_{ij} = 0$ without specifying the values of μ_i. Nevertheless, I was able to obtain explicit maximum likelihood estimates for both models, with and without α_{ij} fixed. This is not always the case. For example, if I want to set the interaction, γ_{jk}, in Equation (2.6) to zero, yielding the independence described by Equation (2.13), I can no longer obtain explicit estimates for the other parameters, and, hence, a value for the normed likelihood. Generally, in Chapter 2, exact explicit estimates were only available in saturated models, where the number of unknown parameters equalled the number of equations to solve (see Section 3.3.1 below). In other cases, some iterative numerical technique of successive approximations, by computer, must be used.

However, we can easily calculate deviances for the complete elimination of one or both explanatory variables in a three-way table. Thus, if one variable, say that indexed by j, is eliminated, the negative log normed likelihood will be

$$\sum_i \sum_j \sum_k n_{ijk} \log(n_{ijk}) + \sum_k n_{\bullet\bullet k} \log(n_{\bullet\bullet k}) - \sum_i \sum_k n_{i\bullet k} \log(n_{i\bullet k})$$
$$- \sum_j \sum_k n_{\bullet jk} \log(n_{\bullet jk})$$

This corresponds to the independence of Equation (2.14). If the response is independent of both variables, as in Equation (2.15), the negative log likelihood will be

$$\sum_i \sum_j \sum_k n_{ijk} \log(n_{ijk}) + n_{\bullet\bullet\bullet} \log(n_{\bullet\bullet\bullet}) - \sum_i n_{i\bullet\bullet} \log(n_{i\bullet\bullet})$$
$$- \sum_j \sum_k n_{\bullet jk} \log(n_{\bullet jk})$$

Example (continued)
For the data on listening to classical music in Table 2.11, the model with complete independence from age has a negative log normed likelihood of 73.9 with a difference of two parameters, that for complete independence from education has 26.1 also with a difference of two parameters. The negative log likelihood, when listening depends on neither variable, is 74.1 with a difference of three parameters. These values are all so large, with a small change in number of parameters, that independence is clearly rejected in every cases.   □

As we saw in the analysis of this example in Section 2.2.4, it does not make sense to set just any parameter to zero. The hierarchical nature of the model should be retained.

Example (continued)
For the classical music study, the main effect of age should not be set to zero because its interaction with education is required in the model.   □

From these examples, we also see that we need some objective way of deciding when negative log normed likelihoods are large (poor fit), especially when the differences in the number of parameters (the model complexity) change.

3.4 Calibrating the likelihood

Statisticians have attempted, in a number of ways, to resolve this problem of weighing goodness-of-fit against complexity. I shall now look briefly at the most important ones. The choice generally depends on the goals of the data analysis, as outlined in Section 3.1.

3.4.1 DEGREES OF FREEDOM

For any set of data, there are usually many possible models that could be fitted to them.

Example
For a $2 \times J$ frequency table, assuming a null logistic model with the same probability for all J categories will usually yield a model that fits poorly. On the other hand, assuming a saturated model with different possible probability for each cell gives a model that fits perfectly. The first model has only one parameter and the second has many. When the explanatory variable is quantitative, we saw in Section 2.2.5, that an intermediate model is logistic regression with two parameters.□

The more parameters used in a model, the more chance there is that it will fit well. Note that only the parameters estimated from the data allow the model to fit more closely. Parameters fixed at some value, independently of the data observed, do not. (Of course, if such a parameter is changed to zero, so that it disappears from the model, the latter may fit more or less well as compared with that with the fixed parameter.)

When comparing two models, as in a likelihood ratio or deviance, it is important to note the number of parameters estimated (calculated from the data) in each. The difference in these numbers between models is called the (difference in) *degrees of freedom* (d.f.).

Example
When we fix $\alpha_{ij} = 0$ in a two-way table and compare this with a model where they are not fixed, the number of degrees of freedom is $(I-1) \times (J-1)$, because the parameters sum to zero over both indices (or have some equivalent constraint such as the baseline). We already saw this from a different point of view in Section 2.1.1.   □

One might think that a model that fits the data more closely would always be better. However, this is not generally the case. We must remember that we have a sample from the population.

- We are interested in the underlying data generating mechanism. A simpler model is usually easier to understand: for example, if interactions are small, we leave them out. A smooth curve, defined by few parameters, is easier to understand than a bumpy one requiring many parameters.
- If the model represents the sample too well, it will have little or no chance of representing a second, similarly generated, sample very well. A model too close to a sample will usually be too far from the population.
- If two models appear to be equally good at explaining a phenomenon, scientists will generally choose the simpler one.

Thus, we want to balance goodness-of-fit, as measured by a large normed likelihood or small deviance, and a simple model, as measured by a large number of unused degrees of freedom, indicating that further parameters could have been added to the model. This is a problem that can generally only be resolved by scientific judgement in each particular case.

3.4.2 MODEL SELECTION CRITERIA

In scientific model construction, the goal is usually to compare the competing models and find the simplest model that adequately describes the data. Then, a model selection criterion is appropriate.

The most widely used such criterion is called the *Akaike information criterion* (AIC). A well fitting model will have a relatively small negative log (normed) likelihood. However, such small values need to be increased in some way for more complex models in order to penalise them. The AIC does this by simply adding the number (p) of estimated parameters to the negative log normed likelihood:

$$\text{AIC} = -\log(\text{R}) + p$$

Because the AIC is only used to *compare* models, any constant can be added to it without changing the results. Thus, for example, R can be replaced by the likelihood L.

The original definition of the AIC used the deviance (D) to which was added two times the number of parameters estimated, $D + 2p$, so that it is twice as large as that used here. In other words, here I shall use that based on the negative log likelihood, not the deviance. In the literature, care should always be taken to ascertain which is used in a particular context.

For a given data set, models with smaller AICs will be preferred. Thus, for example, if several models have similar *relatively* smaller values, they should all be seriously considered. Note, however, that the actual size of any given AIC is arbitrary because any fixed constant could be added to it. The size of the AIC will also depend on the sample size, larger samples generally having relatively larger AICs. Thus, only the relationships among AICs calculated from exactly the same data are relevant.

The AIC, and other such model selection criteria, use the likelihood function for a given model function with the maximum likelihood estimates of all unknown parameters inserted. In this way, each model is given its best chance of explaining the observed data. However, models with more unknown parameters intrinsically have a better chance of being close to the data. The penalty for the number of parameters compensates for this.

The AIC can be related to the construction of likelihood intervals described in Section 3.2.3. For one parameter, if such an interval excludes the zero parameter value, this indicates that a model with the parameter is preferable to one without it. For the AIC to indicate that the model with the parameter is preferable, the negative log normed likelihood for that model must be (at least) one unit smaller than that without the parameter: this compensates for the unit penalty for the extra parameter. The corresponding normed profile likelihood is $e^{-1} = 0.37$. Thus, the likelihood interval corresponding to the AIC is defined by taking this value for R. In other words, if the model with the parameter set to zero does not fall in the interval defined by $R = 0.37$, the AIC will indicate that the model containing that parameter is preferable (that is, that the parameter is required in the model).

Examples (continued)
For Table 1.3 on graduate study plans, the AIC for the independence model, with one parameter, is $0.88 + 1 = 1.88$, whereas that for dependence, with two parameters, has zero log normed likelihood, so that the AIC is $0 + 2 = 2$. Thus, the independence model is indicated as slightly preferable, although there is not much difference between them.

For the malformation data in Table 2.14, the AIC for the logistic regression model is 2.97 with two parameters. When the slope is set to zero, the AIC is 4.10 with one parameter, indicating a poorer model. This can be seen in Figure 3.4 where the value $\beta_1 = 0$ does not lie in the 37% likelihood interval (between the dotted vertical lines).

If the deviance is used, all these AIC values are doubled, but the conclusions are unchanged.   □

The above procedure usually gives satisfactory results, especially if an appropriately sized sample was collected (Section 3.6). However, it may sometimes be desirable to modify the penalty added to the negative log likelihood so that it is a multiple of the number of estimated parameters. The larger this multiple is, the simpler will be the models obtained. Of course, one should decide on the multiple to use before collecting the data. Changing the multiple after seeing the data opens one to the accusation of modifying the procedures in order to obtain the results desired!

One alternative to the AIC that uses a different multiple is called the *Bayesian information criterion* (BIC). This is given by

$$\text{BIC} = -\log(R) + p\log(n)/2$$

where n is the sample size. Instead, it is also often based on the deviance: $D + p\log(n)$. One major disadvantage of this criterion is that it depends on the sample size, so that sample size calculations (Section 3.6) based on it cannot be made before beginning a study.

3.4.3   SIGNIFICANCE TESTS

In a decision-making situation, we must assume that some model is the true one upon which we should base our actions. (In a scientific context, there is never a true model.) Then, the classical statistical approach is to use *frequentist* procedures.

Examples
For a binomial distribution, the particular model that interests us might be that with equal probability of the two events occurring, that is $\pi_1 = \pi_2 = 0.5$.

In a log linear model for a two-way contingency table, it might be independence: $\alpha_{ij} = 0$ for all i and j.   □

Table 3.3. Deviances, and their probabilities (from Table 3.1), for a sample of ten from a binomial distribution with $\pi_1 = 0.5$.

n_1	Probability	Deviance
0	0.001	13.86
1	0.010	7.36
2	0.044	3.85
3	0.117	1.65
4	0.205	0.40
5	0.246	0.00
6	0.205	0.40
7	0.117	1.65
8	0.044	3.85
9	0.010	7.36
10	0.001	13.86

In this section, I shall work with the deviance instead of the negative log normed likelihood, for reasons that will soon become clear. Then, given the true model, the probabilities of occurrence of various sizes of deviances in possible samples can be calculated. However, it is important to realise that a 'true' model must never be deduced from looking at the data. It must be specified before the data are collected. Otherwise, the probabilities calculated for the deviances will be meaningless.

Example
Let us assume that a referendum is to be held and that the government wishes to decide whether or not further publicity for its point of view is required. This will be so if one half of the people in the population would say yes to the question. Then, the binomial distribution with $\pi_1 = 0.5$ might be a reasonable model.

Suppose that we intend to take a sample of ten future voters. The probabilities of the different possible results were given in the appropriate column of Table 3.1. In each case, we can calculate the corresponding deviance, obtained using Equations (3.2) and (3.3). These values are given in Table 3.3. We see that the larger the deviance, the more rarely will it occur.   □

This example indicates a general result. In the same way that sample relative frequencies vastly different from the population probability are rare, large deviances will also be rare if our assumed, fixed model is correct.

Notice that Table 3.3 does not represent a deviance function, which would be for *one* given set of data for *different* models. Instead, it gives the deviances for one model for different possible data sets, only one of which will be observed. This is a probability, not a likelihood, calculation; it can always be carried out before obtaining the data.

Unfortunately, we would have to calculate a different table for every fixed model that we might contemplate and for every sample size that we might choose. However, if the number of observations is not too small, these probabilities can be reasonably approximated by a probability distribution *for any true model* of the population. The appropriate distribution for a deviance is called the Chi-squared (χ^2) distribution. (This explains why the negative log likelihood is multiplied by two to yield the deviance.)

The deviance is, thus, an approximate χ^2 statistic. The shape of the χ^2 distribution depends only on the number of degrees of freedom. Some values are given in Table A.1 in the Appendix. Again, we see that large values are rare. (Two examples of such a distribution are plotted in Figure 4.29. These indicate a potential problem with this approach: for large degrees of freedom, *small* values, that is, observations close to the correct model, are rare!)

Usually, the probability of the observed deviance is not used. Instead, the total probability of all deviances at least as large as that observed is calculated.

Example
Let us calculate the probability of a deviance at least as large as 3.85 for our binomial model with $\pi_1 = 0.5$. This is obtained by adding the three smallest probabilities in each tail in Table 3.3. The sum of the probabilities for $n_1 = 0, 1, 2, 8, 9, 10$ is 0.11. From the first line of Table A.1, the probability of a χ^2 value as large as 3.84 is 0.05. Here, the approximation is poor, because we only have ten observations and there are large jumps between successive possible tail probabilities.   □

Thus, we can determine, at least approximately, whether or not a deviance that we have calculated is rare or unusual for any model that we might hypothesise to be true. If the deviance is large, and the corresponding probability is small, only one of two things could have happened:

(1) a rare event has occurred or
(2) the hypothesised model is wrong.

In such a situation, we would tend to reject the hypothesised model. This is called a *significance* or *hypothesis test* and the exact (or approximate χ^2) probability is called a *P-value*. If such a technique is to be used, both the deviance and the *P*-value should be reported. Several typical values are listed in Table 3.4.

Obviously, things are not black and white. Often, it may be reasonable to reject a model if the *P*-value is, say, less than 0.02 or 0.01. If it is greater than 0.10, we may conclude that the data provide little evidence against the hypothesised model, but they also provide little evidence against many other models! We cannot, then, simply accept the hypothesised one; this is not a model selection or model comparison procedure. For intermediate values, we are uncertain and require further data. This is the question of sample size that I shall discuss in Section 3.6.

A significance test differs logically from the likelihood and deviance from

Table 3.4. Normed likelihoods, deviances, and P-values for one degree of freedom obtained using the χ^2 approximation.

Normed likelihood	Deviance	P-value
0.60	1.0	0.312
0.30	2.4	0.121
0.25	2.8	0.096
0.20	3.2	0.073
0.15	3.8	0.051
0.10	4.6	0.032
0.05	6.0	0.014
0.01	9.2	0.002

which it is derived. The latter two are calculated solely from the models under consideration and the *observed* data. The P-value also uses probabilities of deviances for *data that were not observed*, those that are rarer than that observed. This has given rise to widespread criticism.

Use in model selection   Significance tests have been widely misused as a scientific model selection tool. It is informative to compare their operating characteristics with that of proper model selection criteria (Section 3.4.2). From Table A.1, for one parameter (one degree of freedom), such a test, using a 5% significance level, is equivalent to a penalty of $3.84/(2 \times 1) = 1.96$ in the AIC, for two parameters to $5.99/(2 \times 2) = 1.50$ per parameter, for three parameters to $7.82/(2 \times 3) = 1.30$, and so on. For seven parameters, it is about one, the same as the AIC; thereafter, it becomes increasingly smaller than one. Thus, the penalty changes with the number of parameters, being smaller per parameter for more complex models. Parameters in complex models are not penalised as much as those in simple models.

Another important disadvantage of significance tests is that the true model must be a special case, a simplification, of the alternative available; it is said to be 'nested' in the alternative. Thus, such tests cannot easily be used to compare functionally different models. Model selection criteria do not have this handicap.

The application of significance tests to scientific model selection is also inappropriate for several other reasons:

(1) To call a scientific model true is a contradiction. Any model, by definition, is an approximation to reality, not an exact representation of it.

(2) If a series of different models is to be examined, they usually cannot all be completely specified beforehand. Thus, the 'true' model for each successive test will illegitimately be obtained from the data.

(3) An hypothesised 'true' model will not be rejected unless a better alternative is available.

(4) No definitive decision among a finite set of known alternatives is to be made.

(5) We want to make scientific judgements about models based on the observed data, not on data that might have been observed, those with larger and thus rarer deviances.

Example (continued)

For Table 1.3 on graduate study plans, we found above that the deviance was 1.77. A 2×2 table has 1 d.f., so that, from Table A.1, a value at least as large as this would occur between 10% and 20% of the time, a rather common happening (computer calculation gives a P-value of 0.183). We can conclude that there is no evidence of lack of independence between sex and plans.

Although the hypothesis of independence can easily be specified before collecting the data, it certainly cannot be imagined to be exactly true. As well, no decision is specified, given the outcome of the test.   □

Thus, when selecting among different models for the same data, calibrating by a series of significance tests proves to be unsatisfactory. This is especially so when large numbers of parameters are involved. However, these criticisms of the use of significance tests as a scientific model selection criterion do not detract from their importance in decision making.

Confidence intervals   We saw above in Section 3.4.2 that the AIC implicitly specifies a likelihood interval. The same is true if a fixed P-value is used. Then, the likelihood interval is known as a *confidence interval*. As with significance tests, it is not valid to construct such intervals for parameters in a model selected using the data.

The significance level refers to parameter values excluded from a confidence interval. Thus, from Table 3.4, a 15% likelihood interval corresponds approximately to a 5% significance level, yielding a 95% confidence interval. The interpretation is that, if the model is true, 95% of such intervals constructed (from different data sets) will contain the true parameter value.

Example (continued)

We can calibrate in this way the profile likelihood curve for the malformation data of Table 2.14, given in Figure 3.4. A normed likelihood of $\exp(-3.84/2) = 0.141$ corresponds to deviance of 3.84. From Table A.1, a χ^2 value of this size yields a 5% P-value and hence a 95% confidence interval. For this parameter, the interval for which the normed profile likelihood is at least this large is $(-0.52, -0.01)$ as shown by the vertical solid lines on the graph.

Often, an approximate interval is calculated based on the standard error. For the slope parameter $\widehat{\beta}_1 = -0.317$, the standard error is 0.125. Using the procedure outlined in Section 3.2.4, we obtain the approximate profile likelihood curve shown as the dashed line in Figure 3.4. Note that this curve is always symmetric about the maximum likelihood estimate whereas the exact one generally is not. The corresponding approximate 95% confidence interval is also shown on the graph. Here, the interval is $(-0.56, -0.07)$. This approximation incorrectly (according to the 95% confidence criterion) appears to provide stronger evidence that the interval does not include zero. $\square$

Care must be taken not to mix inferences using model selection criteria (Section 3.4.2) with those using other approaches to inference. It is not valid to choose a model using such criteria and then to apply confidence intervals to its parameters because the model has been chosen based on the data. As well, contradictions can easily arise. The AIC, corresponding to a likelihood interval based on R = 0.37, may indicate that a parameter is required in a model (that is, that it should not be set to zero) whereas the wider 95% confidence interval, corresponding to R = 0.14, subsequently calculated for that parameter, may contain zero.

Example (continued)
The dotted lines in Figure 3.4 define the interval of precision corresponding to the use of the AIC for the slope parameter of the malformation data. We see that it is much narrower than the 95% confidence intervals, thus much more clearly excluding the model with $\beta_1 = 0$. $\square$

Certain scientific publications have begun to require confidence intervals instead of significance tests for important parameters. However, as we have seen, the two are essentially the same. Indeed, if the standard error is used to obtain an approximate interval, the P-value of a significance test transmits exactly the same information as the corresponding interval. It is always preferable to report the likelihood function itself, or at least profile likelihood graphs of all important parameters.

3.4.4 PRIOR PROBABILITY

A third branch of statistical inference, the *Bayesian* approach, is oriented towards procedures for *personal* decision-making. To use these methods, one must have available or be able to obtain numerically quantifiable information about the relative plausibility of *all* possible model functions, and their parameter values, before making observations. This information may then be used to weight the likelihood function by means of Bayes' formula in Equation (1.4).

Suppose that this prior information can be represented as a function of the unknown parameter, say $p(\theta)$. This is called the *prior probability distribution*. Then, from this formula, the new, updated weighting, after making the observations, will be

$$p(\theta|y) = \frac{\Pr(y; \theta)p(\theta)}{\Pr(y)}$$

where $\Pr(y; \theta)$ is just the likelihood function and $\Pr(y) = \sum_\theta \Pr(y; \theta) p(\theta)$. In this way, prior knowledge $p(\theta)$ is combined with the new information in the sample $\Pr(y; \theta)$ to yield a *posterior probability distribution $p(\theta|y)$*.

Example

For our data with 6 yeses in 10 questionnaires, let us look at the simple case where the only models under consideration are $\pi_1 = 0.2$ and $\pi_1 = 0.5$. If we have no prior preference for either value, we would assign the same weight, that is, the same prior probability of being true, to each. Then, $p(\pi_1 = 0.2) = p(\pi_1 = 0.5) = 0.5$. From Table 3.1, $\Pr(n_1 = 6; \pi_1)$ takes the values 0.006 and 0.205, respectively, for the two models for the given data. Then, $\Pr(n_1 = 6) = 0.006 \times 0.5 + 0.205 \times 0.5 = 0.105$. Finally, we obtain the posterior probabilities of the two models:

$$
\begin{aligned}
p(\pi_1 = 0.2 | n_1 = 6) &= \frac{0.006 \times 0.5}{0.105} \\
&= 0.028 \\
p(\pi_1 = 0.5 | n_1 = 6) &= \frac{0.205 \times 0.5}{0.105} \\
&= 0.972
\end{aligned}
$$

Thus, the prior probabilities of one half, indicating indifference between the two models before making the observations, have been modified to posterior probabilities, in the light of the data, which very much favour the model, $\pi_1 = 0.5$. The posterior odds are $0.972/0.028 = 34$ to one in its favour. This is the same as the likelihood ratio calculated for the same two models in the example of Section 3.2.1 above. This only occurs if the prior probabilities are equal.   □

Notice, in Bayes' formula, that any model or parameter value that has zero prior probability will always have zero posterior probability. Thus, scientific *discovery* is excluded by these procedures. As well, Bayesian methods rely primarily on weighted averages of different models. This contrasts with the AIC, which uses maxima (gives the best chance to each model). Such Bayesian averages of various different models are generally very difficult to interpret scientifically.

When an interval of plausible values for a parameter is obtained from a posterior distribution, it is called a *credibility interval*. As we saw in the example, if a flat ('noninformative') prior is used, the posterior distribution is the same as the likelihood function. Then, the credibility intervals will usually be the same as confidence intervals. If there are several parameters, inferences can be made about one of them by averaging out the others using integration. For a flat prior, these will be similar to profile likelihoods. However, these procedures are relatively complex and will not be illustrated here.

If the Bayesian approach of this section is used, the likelihood function should also be reported, because someone, with other prior information, may wish to use a different weighting.

3.5 Goodness-of-fit

In scientific research, it is always useful to have some general indication of how well the selected models actually fit to the data. Both global measures and specific measures for parts of the data can be used. If all available models show indications of poor fit, then better models must be sought.

3.5.1 GLOBAL FIT

When one clearly-defined saturated model exists, comparisons can be made with it. Thus, whenever we make a likelihood comparison of some model of interest with the saturated model, as in the previous sections, we are looking at a global *goodness-of-fit*.

Example (continued)
For the malformation data in Table 2.15, we compared the saturated model with the logistic regression model. We can now examine the goodness-of-fit. The likelihood is just the product of binomial probabilities for each level of alcohol consumption

$$L(\pi) = \prod_j \pi_{1|j}^{n_{1j}} (1 - \pi_{1|j})^{n_{2j}}$$

We can substitute in the two sets of estimates from Table 2.15 and take the ratio of likelihoods so obtained. The negative log likelihood is 0.97 and the AIC 2.97 with $p = 2$ parameters. The negative log normed likelihood for the saturated model is zero with $p = 5$, so that the AIC is 5. Thus, the logistic regression fits well as compared with the saturated model. The decrease in malformed children for mothers with < 1 drinks per day may just be random variation. We also saw above that, when the slope is set to zero (the null model), the AIC is 4.10 with $p = 1$. Thus, based on the AIC, the logistic regression is the preferable of the three models according to these data. □

Probability densities An important situation in which we shall want to examine goodness-of-fit occurs when we have a quantitative variable and wish to simplify the multinomial distribution by using a probability density function as in Equation (1.10). I shall look at a number of such specific density functions in Chapter 4. Such functions will not be able to follow the irregularities observed in a histogram; they *smooth* the empirical distribution in order to isolate the systematic part. Thus, we would like to know whether or not such a function still reasonably represents the data that we have observed.

We know that the maximum likelihood estimates of the multinomial parameters will be $\hat{\pi}_i = n_i/n_\bullet$. The density function, say $f(y_i; \theta)$, will have one or more unknown parameters, such as the mean. These can also be estimated by maximum likelihood to give $\hat{\theta}$. However, from Equation (1.10), these will yield a different set of estimates for the multinomial probabilities. I shall call these $\tilde{\pi}_i$:

$$\tilde{\pi}_i = f(y_i;\hat{\theta})\Delta_i$$

These alternative estimates of π_i can be substituted into the normed multinomial likelihood, the generalisation of Equation (3.2), to give

$$L[f(y_i;\hat{\theta}), \text{ all } i] = \prod_i \left(\frac{\tilde{\pi}_i}{\hat{\pi}_i}\right)^{n_i}$$

$$= \prod_i \left[\frac{f(y_i;\hat{\theta})\Delta_i}{\hat{\pi}_i}\right]^{n_i} \tag{3.7}$$

This provides a measure of goodness-of-fit of a specific probability density function that can be examined using the AIC. It compares the best fit under the smoothing constraints of the density function with the best overall fit of the multinomial distribution.

The multinomial model has $I - 1$ unknown parameters, given only the constraint that the probabilities sum to unity. The density function usually will have only one or two parameters, often yielding a very substantial reduction in complexity over the multinomial. The difference between these two numbers gives the number of degrees of freedom.

In the appropriate context, a significance test can be applied to the deviance. If so, and the χ^2 approximation is used, care should be taken that categories are constructed such that the frequencies are not too small, say all greater than 5.

3.5.2   RESIDUALS AND DIAGNOSTICS

Goodness-of-fit comparisons using the likelihood function only provide a global measure of the distance between a model of interest and the model fitting the data exactly. If the model does not fit well, they do not usually indicate why. If no theory is available to help to construct a better model, we need a set of diagnostic procedures to determine the reasons for lack of fit.

The goal of a model is to describe the systematic variation in the data. Once the model takes this variation into account, only random variation should be left. The usual way to examine this is by calculating the *residual* for each observation. Because residuals are supposed to represent random variation, they should contain no interpretable patterns.

A (raw) residual is usually defined as a difference between an observation and its fitted value, that is, the value predicted for it by the model. However, these differences can be very variable in size. Thus, they generally need to be corrected by dividing by their standard deviations.

An extreme observation, *with respect to some given model*, is called an *outlier*. An outlier may arise in at least three distinct ways:

(1) An error has been made in measuring or recording the observation.

Table 3.5. Residuals for the independence model applied to the data on graduate study plans of Table 1.3.

	Log linear		Logistic
	PhD plans		
Sex	Yes	No	
Male	−0.650	0.523	−0.834
Female	0.808	−0.650	1.036

(2) Some rare event has occurred.

(3) The model is not adequate to describe the observations.

Inspection of residuals may allow us to determine whether or not certain observations are outliers. Naturally, the first thing to check is whether the data are correct or not. If no problem is found, then one must decide whether or not the discrepancies merit developing an improved model.

Residuals for log linear models   Consider, first, residuals for a log linear model. We saw in Section 2.1.2 that the fitted values for the independence model in a two-way contingency table are $n_{\bullet\bullet}\widehat{\pi}_{i\bullet}\widehat{\pi}_{\bullet j}$. Then, the *standardised residuals* for a contingency table are defined as the observed frequency minus the fitted value all divided by the square root of the fitted value:

$$\frac{n_{ij} - n_{\bullet\bullet}\widehat{\pi}_{i\bullet}\widehat{\pi}_{\bullet j}}{\sqrt{n_{\bullet\bullet}\widehat{\pi}_{i\bullet}\widehat{\pi}_{\bullet j}}}$$

The denominator is the estimated standard deviation of the observed frequency. (We shall see in Section 4.3.1 how it is obtained.)

Example (continued)
Let us look once more at the contingency table for graduate study plans, originally given in Table 1.3. We wish to examine in more detail the model for independence between sex and plans. In fact, we calculated the fitted values for this model in Table 2.5. For these data, the residuals are given in the first (log linear) part of Table 3.5. For example, the upper left value is calculated as $(5 - 6.68)/\sqrt{6.68} = -0.650$. We see that the independence model under-predicts 'no PhD plans' for males and 'PhD plans' for females, as might be expected. However, residuals are not very informative in a small table, such as this, with only one degree of freedom. □

Residuals for logistic models   When a logistic model is fitted, it is traditional only to look at the residuals for the first category of response. We saw in Section 1.3.4 that the variance for n_1 having the binomial distribution is $n_{\bullet}\pi_1(1 - \pi_1)$.

Table 3.6. Residuals for the independence model applied to the data on birth defects of Table 2.14.

| Alcohol | Log linear | | Logistic |
| | Malformation | | |
consumption	Absent	Present	
0	−0.030	0.600	−0.601
< 1	0.047	−0.859	0.860
1–2	−0.061	0.954	−0.956
3–5	0.013	−0.141	0.142
≥ 6	−0.020	0.130	−0.132

The expected value for this category is $n_{\bullet}\pi_1$. Then, the residuals for a model with one explanatory variable are given by

$$\frac{n_{1j} - n_{\bullet j}\widehat{\pi_{1j}}}{\sqrt{n_{\bullet j}\widehat{\pi_{1j}}(1 - \widehat{\pi_{1j}})}}$$

Of course, by changing the first index to 2, residuals for the second category could also be calculated but they will be identical except for a change of sign.

Examples
The residuals for the data on study plans, calculated in this way, are also shown in Table 3.5. We can see that they are considerably larger. This is because of the extra factor of $(1 - \widehat{\pi_{1j}})$ in the variance in the denominator. The difference can be interpreted as arising because the log linear model uses a joint distribution of the two variables whereas the logistic model assumes a conditional distribution of the response, depending on sex.

Let us now look at the residuals for the logistic regression model fitted to the malformation data in Section 2.2.5. These are shown, calculated in the two ways, in Table 3.6. We see that those calculated for the logistic model (for the first response category) are almost identical to the ones for a log linear model for the second category (except for the change of sign). This arises because the first category of response has very large frequencies as compared with the second. The observed value of malformation absent with < 1 drink consumption larger than the fitted value in Table 2.15 is reflected in the positive residual here. However, it is compensated by the negative residual in the next category, 1–2 drinks.

Both these models fit well, so that we do not expect to find any informative pattern in the residuals. In Chapter 4, we shall encounter examples with patterned residuals.   □

Examination of residuals is most useful when the model under study is not too close to the saturated model, that is, when the number of degrees of freedom is reasonably large. Especially if some of the variables are continuous, it is often useful to plot the residuals against these variables to look for patterns. These may indicate ways in which the model could be improved.

Components of goodness-of-fit If we sum the squares of the standardised residuals in a contingency table, we obtain another global measure of goodness-of-fit, an approximation to the deviance, called the *Pearson statistic*. For large enough sample size, this also can be approximated by a χ^2 distribution for making a test of significance. Thus, the Pearson statistic should have a value close to the deviance.

Examples
From the log linear residuals in Table 3.5, we can calculate the Pearson statistic:

$$(-0.650)^2 + 0.523^2 + 0.808^2 + (-0.650)^2 = 1.77$$

For this table, this happens to be identical to the deviance.

For the logistic regression applied to the malformation data, the deviance is 1.95. From Table 3.6, the Pearson statistic is

$$(-0.601)^2 + 0.860^2 + (-0.956)^2 + 0.142^2 + (-0.132)^2 = 2.05$$

Here there is a small difference. □

The procedure just outlined for obtaining an approximate global measure of goodness-of-fit implies that we could proceed in the reverse order with the deviance. We can decompose it into its separate components and examine them. As we have seen, for independent observations, a deviance is calculated as a sum, with one term for each observation. These terms may be positive or negative. We can take the square root of the absolute value of each term and then put back the positive or negative sign. These are called *deviance residuals*. They can be examined in the same way as the more classical residuals based on fitted values described above.

3.6 Sample size calculation

In designing a study, it is usually very important to know how large a sample will be required in order to be able to detect a difference in distribution of some specified size among the subgroups. For example, we are often interested in comparing a model having no difference with one having some prespecified difference among the subgroups. The ability of an inference procedure to detect such a difference is called its *power*.

To make a sample size calculation, we must be able to specify completely the parameter values of the two models to be compared. In other words, we must assume known that which we set out to learn from the study!

Then, with these fixed values in the model function, the normed likelihood will only be a function of the sample size $n_\bullet$ and of the maximum likelihood estimate. Because we wish to distinguish between two models, the worst case situation that could happen would be when the estimate is centred between them. In such a case, both would be equally likely by the criterion we choose. Then, we

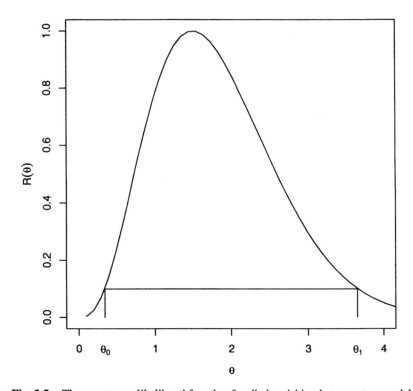

Fig. 3.5. The worst case likelihood function for distinguishing between two models.

want each of them to be just at the limit of being unlikely. If both are unlikely by our criterion, the sample size is too large; if both are likely, it is too small.

Suppose that we wish to be able to detect a specified difference between two values of some parameter, θ, say between θ_0 and θ_1. This situation is illustrated in Figure 3.5, where a normed likelihood function, similar to Figure 3.1, is plotted. However, here the curve does not arise from observed data, but is chosen to lie between the two values of interest, θ_0 and θ_1, with equal likelihood of say 10% for each of them, the chosen criterion. If the situation illustrated in this graph were to arise, we would have no preference between θ_0 and θ_1 because they would be equally likely, given the data.

If $\hat{\theta}$ turns out to be different from that indicated by the curve, as it almost surely will, the curve will move either to the left or to the right. Then, depending on which way it moves, either θ_0 or θ_1 will have a larger likelihood than 10% and the other smaller so that we can choose between them. Thus, Figure 3.5 shows the worst case situation when we cannot distinguish between θ_0 and θ_1.

Example

Suppose that we want to plan a study to see if there is a difference between the two sexes in the response to a yes/no question. A reasonable model will involve the difference between two binomial distributions, as described by a 2×2 contingency table. We shall use the logistic model of Equation (2.4). We shall choose the sample of as yet unknown size $n_\bullet$ in such a way that the explanatory variable, sex, has equal numbers, $n_\bullet/2$, in each category.

Suppose that we wish to detect a difference in response distribution between the sexes corresponding to an odds ratio of about 5. This is a log odds ratio of about 1.6. From Equation (2.5), this log odds ratio corresponds to $\alpha_1 = 0.8$ in the logistic model with the mean constraints. Because we wish to distinguish between $\alpha_1 = 0$ and $\alpha_1 = 0.8$, the worse case will occur if we obtain about $\widehat{\alpha_1} = 0.4$. Finally, we need to know the mean probability of response for the two sexes. Suppose initially that this is 0.5 so that $\mu = 0$. Then, in the worst case sample, the number of the first sex saying yes will be

$$\frac{n_\bullet}{2}\widehat{\pi_{1|1}} = \frac{n_\bullet}{2}\frac{e^{\widehat{\alpha_1}}}{(1+e^{\widehat{\alpha_1}})}$$
$$= \frac{n_\bullet}{2}\frac{e^{0.4}}{(1+e^{0.4})}$$
$$= \frac{n_\bullet}{2}0.599$$

whereas, for the second sex, it will be

$$\frac{n_\bullet}{2}\widehat{\pi_{1|2}} = \frac{n_\bullet}{2}\frac{e^{\widehat{\alpha_2}}}{(1+e^{\widehat{\alpha_2}})}$$
$$= \frac{n_\bullet}{2}\frac{e^{-0.4}}{(1+e^{-0.4})}$$
$$= \frac{n_\bullet}{2}0.401$$

For the comparison of the model having fixed $\alpha_1 = 0$ or $\alpha_1 = 0.8$ with the best model for these 'observations', there is a difference of one parameter because the fixed parameter value is not estimated from the data. If we use the AIC, we are accepting that a difference in log likelihood of one is sufficient for two models, with a difference of one estimated parameter, to fit equally well. (This is equivalent to a likelihood ratio of $e^{-1} = 0.37$.) From the normed binomial likelihood of Equation (3.5), with $\pi_{1\bullet} = 0.5$, we can obtain the negative log likelihood ratio:

$$-\log[R(\alpha_1 = 0 \text{ vs } \alpha_1 = 0.8)] = -n_\bullet\log(0.5) + 0.599n_\bullet\log(0.599)/2$$
$$+ 0.401n_\bullet\log(0.401)/2$$
$$+ 0.401n_\bullet\log(0.401)/2$$
$$+ 0.599n_\bullet\log(0.599)/2$$

We see that $n_\bullet$ is a common factor on the right-hand side. Then, setting this equation equal to one (from the AIC) and solving, we find

$$n_\bullet = \frac{1}{-\log(0.5) + 0.599\log(0.599) + 0.401\log(0.401)}$$
$$= 50.68$$

We shall require a sample of minimum size 51 in order to be able to have, at worst, the same AIC for each of the two models $\alpha_1 = 0$ and $\alpha_1 = 0.8$. All other possible samples of this size will have an AIC for one of the two models greater than the other.

Because, before collecting the sample, we do not know the mean probability of response for the two sexes, we should try a number of different values besides 0.5 or $\mu = 0$. Let us look at a mean probability of 0.75 of saying yes. We shall have $\mu = \log(0.75/0.25) = 1.099$. Then, the numbers saying yes for the two sexes will be

$$\frac{n_\bullet}{2}\widehat{\pi_{1|1}} = \frac{n_\bullet}{2}\frac{e^{1.099+0.4}}{(1+e^{1.099+0.4})}$$
$$= \frac{n_\bullet}{2}0.817$$
$$\frac{n_\bullet}{2}\widehat{\pi_{1|2}} = \frac{n_\bullet}{2}\frac{e^{1.099-0.4}}{(1+e^{1.099-0.4})}$$
$$= \frac{n_\bullet}{2}0.668$$

This means that the total number saying yes is $0.817n_\bullet/2 + 0.668n_\bullet/2 = 0.743n_\bullet$ whereas above it was $0.5n_\bullet$.

The sample size calculations are, then,

$$\begin{aligned}
-\log[R(\alpha_1 = 0 \text{ vs } \alpha_1 = 0.8)] = &-0.743n_\bullet\log(0.743) - 0.257n_\bullet\log(0.257)\\
&+ 0.817n_\bullet\log(0.817)/2\\
&+ 0.183n_\bullet\log(0.183)/2\\
&+ 0.668n_\bullet\log(0.668)/2\\
&+ 0.332n_\bullet\log(0.332)/2
\end{aligned}$$

Setting this equal to one and solving, we obtain $n_\bullet = 70.69$. Here, we shall require a minimum sample size of 71 to detect the desired difference. As the mean probability of response moves away from 0.5, the sample size required to detect that a given log odds ratio is different from zero increases.

If a χ^2 test with a given size of P-value is to be used instead of the AIC, the final expression to be solved for $n_\bullet$ above is set to one-half of the corresponding deviance value instead of to one.   ☐

The same procedure may be used in more complex cases (see also Sections 4.7.2 and 5.5). However, as the number of parameters in the model increases, the number of assumptions that must be made will also increase, making sample size calculations more difficult.

3.7 Exercises

(1) The Poisson distribution (Section 4.3.1 below) is given by

$$\Pr(y_i) = \frac{e^{-\mu}\mu^{y_i}}{y_i!} \qquad y_i = 0,1,2,\ldots$$

The maximum likelihood estimate of the mean is $\hat{\mu} = \frac{1}{n_\bullet}\Sigma_i y_i$. Suppose that this is calculated to be $\hat{\mu} = 10$ with $n_\bullet = 20$ observations.
   (a) Plot the normed likelihood function.
   (b) Repeat for the same estimate but $n_\bullet = 50$ and $n_\bullet = 100$.

(2) Calculate the AICs under independence for the logistic models that were fitted to the following data in the Exercises of Chapter 2:
   (a) injuries in car accidents and wearing a seat belt of Table 1.9;
   (b) myocardial infarction and contraceptive use of Table 1.10;
   (c) opinions about the death penalty and gun registration of Table 2.3.
   In each case,
   (a) plot the normed likelihood function for the dependence parameter and choose an interval of precision;
   (b) discuss the conclusions that can be drawn;
   (c) note whether they change anything that you concluded in Chapter 2.
   Can we conclude
   (a) that making seat belts compulsory will reduce the fatal accident rate?
   (b) that using contraceptives is a cause of myocardial infarction?

(3) Calculate the AICs under independence for the logistic models that were fitted to the following data in the Exercises of Chapter 2:
   (a) the data on soloists' choice of piano in Exercise (2.4);
   (b) the British migration data of Table 1.8.
   In each case,
   (a) discuss the conclusions that can be drawn;
   (b) note whether they change anything that you concluded in Chapter 2.

(4) Calculate the AICs for independence of the response from the explanatory variables for the logistic or log linear models that were fitted to the following data in Chapter 2:
   (a) the tables of Exercise (2.6) concerning listening to the radio;
   (b) the delinquency data of Exercise (2.7);
   (c) the data on factors influencing knowledge of cancer in Tables 2.10 and 2.13 and Exercise (2.8);
   (d) the shopping data of Exercise (2.9).
   In each case,

    (a) select one or more parameters in which you are especially interested, plot their normed profile likelihood function(s), and choose appropriate intervals of precision;

    (b) discuss the conclusions what can be drawn;

    (c) note whether they change anything that you concluded in Chapter 2.

(5) As in the previous question, calculate the AICs for independence of the response from the explanatory variables for the polytomous logistic model for the following data from Chapter 2:

    (a) the political ideology data of Exercise (2.14);

    (b) the media rating data of Exercise (2.15).

In each case,

    (a) redo the AIC calculations for the reconstructed table for the two ordinal variable models;

    (b) compare the results and discuss the meaning of any differences.

(6) Suppose that the shopping data of Exercise (2.9) were collected by a firm considering the construction of a new shopping centre in the same region.

    (a) Specify an appropriate null hypothesis for making such a decision. (Try to do this without using what you already know about these data!)

    (b) Calculate the corresponding test of significance.

    (c) Describe what subsequent action you would advise should be taken.

(7) Suppose that the study on myocardial infarction and contraceptive use reported Table 1.10 was conducted to decide whether or not to withdraw contraceptives from the market.

    (a) Based on the normed likelihood function that you plotted in Exercise (3.2) above, calculate an appropriate confidence interval for the dependence of myocardial infarction on contraceptive use.

    (b) What recommendations would you make to the policy deciders?

    (c) Suppose now that you believe that only two values of the dependence parameter could reasonably be true, one of them being that for independence and the other being a log odds ratio of one.

        i. Assign your prior probabilities to these two possibilities. (Try to do this without using what you already know about these data!)

        ii. Obtain the updated posterior probabilities.

        iii. Has your opinion on the subject now changed and, if so, in what way?

(8) (a) What size of sample would be required to detect that a log odds ratio was 1.0 as opposed to zero? Assume, as in the example in Section 3.6, that you can choose a sample with equal numbers in each category of the explanatory variable.

    (b) Plot the required sample size for several values of the mean probability of response.

4
Probability distributions

In Chapter 2, we looked at a family of models, the logistic and log linear models, to describe how the shape of a probability distribution changes in different subgroups of a population. However, because the response in these models is nominal, their multinomial distributions (and their histogram representations) can take any form. (See, for example, Figure 1.2.) These types of models are really the only ones possible for independent observations when the response is nominal (although there are various possibilities for linking the probabilities to the explanatory variables instead of the log or logit).

On the other hand, more information is available when the response variable is quantitative. In such cases, more understanding about the data generating mechanism can usually be gained if the distribution is constrained to take some particularly appropriate shape. As we saw in Equation (1.10), this can be accomplished by using some specific probability density function in the construction of models. I shall look at models constructed in this way in some detail in this chapter.

4.1 Constructing probability distributions

Before we examine some of the many distributions available, it is important to see how they can be constructed and to recall how they are related to the multinomial distribution.

4.1.1 MULTINOMIAL DISTRIBUTION

Up until now, we have only been concerned with the multinomial distribution, given in Equation (1.8),

$$\Pr(n_1, n_2, \ldots, n_I) = \binom{n_\bullet}{n_1 n_2 \cdots n_I} \prod_i \pi_i^{n_i}, \qquad \sum_i \pi_i = 1$$

and its special case, the binomial distribution, which was given in Equation (1.9),

$$\Pr(n_1, n_2) = \binom{n_\bullet}{n_1} \pi_1^{n_1} (1 - \pi_1)^{n_2}$$

The multinomial distribution is applicable to any empirically observable response variable as long as the observations are independent. As we have seen, the maximum likelihood estimates of the parameters are given by

$$\widehat{\pi}_i = \frac{n_i}{n_\bullet}$$

The mean number of observations in a category is $n_\bullet \pi_i$, and its estimate is called the fitted value. A measure of the variability, or variance, of this n_i, if we took several samples from the same population, is $n_\bullet \pi_i (1 - \pi_i)$. That for $\widehat{\pi}_i$ is $\pi_i (1 - \pi_i)/n_\bullet$ and its standard error is the square root of this. The standardised residuals are given by

$$\frac{n_i - n_\bullet \widetilde{\pi}_i}{\sqrt{n_\bullet \widetilde{\pi}_i (1 - \widetilde{\pi}_i)}}$$

where $n_\bullet \widetilde{\pi}_i$ are the fitted values for some model of interest. However, those given by Equation (3.8) may also be used, as I shall do in this chapter.

4.1.2 DENSITY FUNCTIONS

If the response variable is strictly nominal, the multinomial is the suitable probability distribution for independent observations. If it is ordinal, we can use this structure among the categories in our models, as I did in Section 2.3.4. (I shall present two other possibilities in Section 4.2.)

However, if the response variable is quantitative, either integral or continuous, we can go much further. As we have seen in Equation (1.10), we may link the multinomial probabilities together through the values of the response variable by using a density function:

$$\pi_i = f(y_i; \theta) \Delta_i$$

There are several important reasons for using an appropriate density function, where possible, in a model.

- A density function smooths and simplifies the rough histogram.
- It reduces the number of unknown parameters.
- Hopefully, it will provide better understanding of the underlying data generating mechanism.

Of course, the simplest such mechanism is just the independent random generation of nominal labels as responses, described by the multinomial distribution. But, when the labels are quantitative measurements, more information is available, and should be used.

Example
Usually labels in some ranges of numbers will be more probable than others. In most cultures, a family of size two or three will be much more probable than one of size 20. An appropriate model should not let the probabilities of different family sizes vary haphazardly as the multinomial distribution with nominal responses does. Instead, it should impose that the probabilities of adjacent values vary smoothly and that those around two or three be largest.   □

Construction In principle, density functions are simple to create; all that is required is some function of the response variable that yields a summable series of non-negative weights. Then, each member of the series is divided by the sum, called the normalising constant, yielding the probability of the corresponding response value. In this way, the new series will sum to one, fulfilling the criteria of a probability distribution.

Example
Suppose that the response variable has I possible values and that we have a set of I equal weights, each one arbitrarily set to unity. Then, their sum, the normalising constant, is also I and the density is

$$f(y_i) = \frac{1}{I}$$

so that all response values have the same probability. This is the uniform distribution to be discussed in Section 4.2.1 below. □

Perhaps surprisingly, the number of response categories need not be finite, as long as the sum of their corresponding weights is calculable. This is primarily useful to simplify the mathematics; observed responses can never have an infinite number of different values.

Example
Let us look at the case of a response variable that can take any non-negative integer value i. Consider the weights $\mu^i/i!$. This infinite series happens to be summable:

$$\sum_{i=0}^{\infty} \frac{\mu^i}{i!} = e^{\mu}$$

By dividing each term of the series by its sum, we obtain a density function

$$f(y_i; \mu) = \frac{\mu^{y_i}}{e^{\mu} y_i!}$$
$$= \frac{e^{-\mu} \mu^{y_i}}{y_i!}$$

with $y_i = i$. We shall see below, in Section 4.3.1, that this is called the Poisson distribution. □

However, construction of a probability density function in this way does not guarantee that it will have a useful interpretation as a model for a data generating mechanism. This requires further knowledge of its theoretical properties. We shall be studying these throughout this chapter. As well, in Section 4.6, I shall present another way of constructing new distributions, by transforming the response variable.

Fitting   When data are available that might be thought to follow some density function or other, we can obtain estimates of the parameters using the likelihood function. We shall also want to look at the goodness-of-fit. As we saw in Section 3.5, the normed multinomial likelihood of Equation (3.7), or the corresponding negative log normed likelihood

$$-\sum_i n_i \log\left(\frac{\tilde{\pi}_i}{\hat{\pi}_i}\right) = -\sum_i n_i \log\left[\frac{f(y_i;\hat{\theta})\Delta_i}{\hat{\pi}_i}\right] \tag{4.1}$$

can be used. The larger is this latter value, the further the distribution of interest is from the estimated multinomial distribution (which fits the data perfectly).

As in Section 3.5.1, in the above expression, I distinguish between the maximum likelihood estimates of the multinomial probabilities $\hat{\pi}_i$ (the relative frequencies of the response categories in the sample) and the estimates of these probabilities

$$\tilde{\pi}_i = f(y_i;\hat{\theta})\Delta_i$$

obtained using some density function with the maximum likelihood estimates of its parameters. Note, also, that calculation of this log likelihood requires sufficient decimals in the estimates of π_i to avoid round-off errors. The smaller estimates of π_i given in the tables below (rounded for presentation reasons) are usually not sufficiently precise for this purpose.

The AIC for a given density function is obtained by adding the number of parameters estimated in that function to the negative log normed likelihood in Equation (4.1). For the multinomial distribution, $\tilde{\pi}_i = \hat{\pi}_i$ so that the left-hand side of Equation (4.1) is always zero. Thus, the AIC for the multinomial distribution is always equal to the number of estimated parameters, that is, to the number of response categories minus one. (These parameter estimates are the relative frequencies; it suffices to know all but one because they sum to unity.) In a frequency table, this number can be somewhat arbitrary, depending on whether or not zero frequencies are present and counted.

The number of degrees of freedom (d.f.) is the difference in the numbers of parameters fitted in the density and in the multinomial distribution, that is, the number of response categories minus one minus the number of parameters in the density function. This tells us how many more parameters could be added to the density before reaching a saturated model (the multinomial).

Mean and variance   In Section 1.2.2, I introduced the mean and variance as descriptive statistics. In certain cases, one or both of these will also be estimates of unknown parameters in the density function, represented respectively by a location parameter μ and a dispersion parameter σ^2. (Note, however, that these symbols do not always represent the mean and variance, although they always indicate location and dispersion, respectively.)

In many distributions, a strict relationship holds between the *theoretical* mean, say μ_T, and the *theoretical* variance, say σ_T^2, even though they may not be explicit parameters in that distribution. Then, the standard error of the mean parameter is generally given by $\sigma_T/\sqrt{n_\bullet}$.

I shall now look successively at some of the most important probability density functions representing mainly distributions for integral and continuous variables. Remember, however, that they are *models*, that is, they are only approximations to the underlying data generating mechanism. In many cases, they may not even be at all realistic. They will, nevertheless, often be useful representations to aid us in understanding what is going on. As well, when they can be shown *not* to fit well to the data, they can indicate clearly that a certain mechanism is *not* operating. This can be important information.

4.2 Distributions for ordinal variables

Let us begin by two simple distributions that do not require the response categories to be quantitative.

4.2.1 UNIFORM DISTRIBUTION

The simplest example of a multinomial distribution has all the probabilities equal. If there are I categories, this means that we have

$$\pi_i = \frac{1}{I}$$

This is called the *uniform distribution*. It is illustrated in Figure 4.1 for five categories. This is a useful distribution for showing that all responses do *not* have the same probability.

This distribution has no unknown parameters that need to be estimated from the data. As well, it does not use the values of the variable, so that it can be applied to any type of variable, including nominal. However, the variable is usually at least ordinal, as we shall see in the example. If the variable is quantitative, the theoretical mean and variance are, respectively

$$\mu_T = \frac{I+1}{2}$$

$$\sigma_T^2 = \frac{I^2-1}{12}$$

Example
Table 4.1 shows the numbers of infants born with a certain illness during each month of a year. In the simplest model, the uniform distribution, the theoretical probabilities are all $1/12 = 0.083$. They are plotted as a density function on the histogram in Figure 4.2.

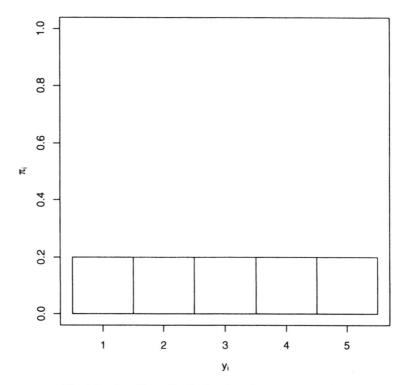

Fig. 4.1. A uniform distribution for a five category variable.

Table 4.1. Numbers of infants born with an illness each month of a year, with the fitted model and the residuals.

Month	Frequency	Fitted	Residual
January	8	9.42	−0.462
February	19	9.42	3.123
March	11	9.42	0.516
April	12	9.42	0.842
May	16	9.42	2.145
June	8	9.42	−0.462
July	7	9.42	−0.788
August	5	9.42	−1.439
September	8	9.42	−0.462
October	3	9.42	−2.091
November	8	9.42	−0.462
December	8	9.42	−0.462
Total	113	113	

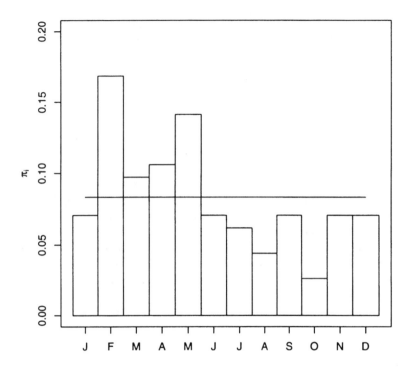

Fig. 4.2. The histogram and uniform density function for the illness of children in Table 4.1.

The negative log normed likelihood is

$$-\left[8\log\left(\frac{1/12}{8/113}\right) + 19\log\left(\frac{1/12}{19/113}\right) + 11\log\left(\frac{1/12}{11/113}\right)\right.$$
$$+ 12\log\left(\frac{1/12}{12/113}\right) + 16\log\left(\frac{1/12}{16/113}\right) + 8\log\left(\frac{1/12}{8/113}\right)$$
$$+ 7\log\left(\frac{1/12}{7/113}\right) + 5\log\left(\frac{1/12}{5/113}\right) + 8\log\left(\frac{1/12}{8/113}\right)$$
$$\left.+ 3\log\left(\frac{1/12}{3/113}\right) + 8\log\left(\frac{1/12}{8/113}\right) + 8\log\left(\frac{1/12}{8/113}\right)\right] = 11.2$$

with $I - 1 - 0 = 11$ d.f. The AIC for this distribution is also 11.2 because no parameters are estimated, whereas that for the multinomial is 11. (As we know, the first should be smaller than the second for the distribution of interest to be plausible.)

Thus, a model of constant probability of the illness for all months of the year is somewhat unlikely. If we look at the histogram, we see that there is much higher probability in the late winter and spring than the other months. Thus, the standardised residuals, as shown in Table 4.1, reveal a pattern, with positive values in these months. This is an invitation to the research worker to find a density function smoothing these data!   □

4.2.2 ZETA DISTRIBUTION

One way in which the categories of a variable may be ordered is by descending order of their frequency, although this can really only be done once the observations are made. Then, the items are said to be ranked. Special distributions will be required in this case. Thus, the zeta distribution was developed in linguistics for the study of word frequencies in texts. However, it can be used much more widely, for example, for the frequency of surnames in a community.

Because the response variable y_i is a rank, the required distribution must have probabilities that uniformly decrease. A simple possibility is to use weights that are reciprocals of the ranks, $1/y_i$. Unfortunately, if we assume an infinite number of possible ranks, the sum of these weights will also be infinite. The denominator must be at least slightly larger than y_i so that enough of the weights are very small. One possibility is to use $1/y_i^\gamma$ with some $\gamma > 1$. The sum of such values,

$$\zeta(\gamma) = \sum_i \frac{1}{y_i^\gamma}$$

is called the zeta function. It is the normalising constant for the infinite series. Then, the zeta distribution will be

$$\pi_i = f(y_i; \gamma)$$
$$= \frac{1}{y_i^\gamma \zeta(\gamma)}, \qquad \gamma > 1, \qquad y_i = 1, 2, 3, \ldots$$

If $\gamma \leq 1$, the series, summed over an infinite number of terms, does not converge to a finite sum. As we shall see in the example, this problem can sometimes be avoided by taking only a small number of terms in the series. This is called a *truncated* distribution.

The zeta distribution has a very long tail. The larger is γ, the faster the probabilities decrease. The distribution for $\gamma = 2$ is plotted in Figure 4.3. The theoretical mean and variance are

$$\mu_T = \frac{\zeta(\gamma - 1)}{\zeta(\gamma)}, \qquad \gamma > 2$$

$$\sigma_T^2 = \frac{\zeta(\gamma - 2)}{\zeta(\gamma)} - \left[\frac{\zeta(\gamma - 1)}{\zeta(\gamma)} \right]^2, \qquad \gamma > 3$$

Thus, this is a distribution that may not even have a mean! The tail may be too thick so that extremely large values are too probable.

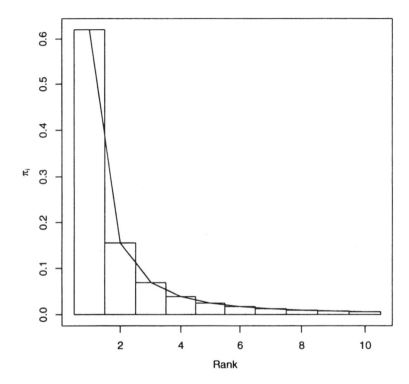

Fig. 4.3. A zeta distribution with $\gamma = 2$.

The maximum likelihood estimate of γ involves the zeta function and can only easily be obtained by using a computer.

Example
Table 4.2 gives the ranks of the total ingot capacity of the top ten USA steel producers in 1958, with their production. Notice that the capacity is not a frequency so that the multinomial distribution is not applicable.

Here, I shall fit a truncated distribution, stopping at ten, instead of going to infinity, because we only have the first ten steel producers. In this way, we can easily calculate the normalising constant, $\sum_{i=1}^{10} 1/y_i^{\gamma}$. This also allows us to set $\gamma = 1$, so that the distribution is simply based on the reciprocals of the ranks. Then, the normalising constant is $\sum_{i=1}^{10} 1/y_i = 2.929$.

The fitted values, $n_{\bullet}\widetilde{\pi}_i$, are also given in the table. The fit seems fairly close, although the two highest ranking companies have their capacity slightly underestimated, and most of the others overestimated. This might be corrected if γ were slightly greater than unity. Indeed, the maximum likelihood estimate is $\hat{\gamma} = 1.15$. The model with this value is plotted in Figure 4.4.

Table 4.2. Steel production capacity (100,000 tons) of the top ten USA producers, with the fitted zeta distribution when $\gamma = 1$ and the residuals. (Derman *et al.*, 1973, p. 299, from Simon and Bonini)

Company	Rank	Capacity	Fitted	Residual $\gamma = 1$	Residual $\hat{\gamma} = 1.15$
US Steel	1	387	339.37	2.586	−0.043
Bethlehem	2	185	169.68	1.176	0.804
Republic	3	103	113.12	−0.952	−0.598
Jones and Laughlin	4	62	84.84	−2.480	−1.852
National	5	60	67.87	−0.956	−0.079
Youngstown	6	55	56.56	−0.208	0.839
Armeo	7	49	48.48	0.075	1.229
Inland	8	47	42.42	0.703	1.979
Colorado Fuel and Iron	9	25	37.71	−2.069	−1.041
Wheeling	10	21	33.94	−2.221	−1.198

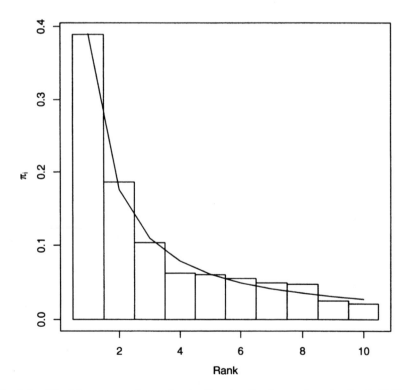

Fig. 4.4. The histogram and zeta density function for the steel production in Table 4.2 with $\hat{\gamma} = 1.15$.

Because the capacities are not frequencies, it does not make sense to calculate AICs for goodness-of-fit. The standardised residuals are presented in Table 4.2 although their denominator is also a problem. We can see that those when the maximum likelihood estimate of γ is used are considerably smaller. There, however, does appear to be some pattern that might be modelled if we had more information about the data.   □

4.3 Distributions for counts

In most cases, integral response variables describe counts of events. Recall (Section 1.2.1) that a count refers to repeated events for the same individual, whereas a frequency refers to independent events for different individuals.

Probability distributions appropriate for integral variables naturally have a unit of measurement, Δ_i, of one, unless some grouping has been applied. Thus, the probability equals the probability density function,

$$\pi_i = f(y_i; \theta)$$

from Equation (1.10). In such cases, it is often simply called the probability (mass) function.

4.3.1 POISSON DISTRIBUTION

Let us consider first the case where only one type of event occurs. For each individual, we count the number of independent events, there being no limit to their total number. Then, the simplest model for these counts will be the Poisson distribution,

$$
\begin{aligned}
\pi_i &= f(y_i; \mu) \\
&= \frac{e^{-\mu} \mu^{y_i}}{y_i!}, \qquad \mu > 0, \qquad y_i = 0, 1, 2, \ldots
\end{aligned}
\tag{4.2}
$$

which we already encountered in Section 4.1.1. In this distribution, μ is the parameter giving the mean number of counts. The histogram and polygon for a typical Poisson distribution are shown in Figure 4.5. Notice that it is asymmetric (*skewed*) with a longer tail to the right. The smaller is the mean, the more accentuated is this form.

The maximum likelihood estimate of the mean is $\hat{\mu} = \bar{y}_\bullet$, as might be expected. To simplify calculations of the probabilities, the recursive relationship

$$
\begin{aligned}
\pi_1 &= e^{-\mu} \\
\pi_i &= \frac{\pi_{i-1}\mu}{y_i}, \qquad i = 2, 3, 4, \ldots
\end{aligned}
$$

can be used. If the last category contains all counts greater than some value, the data are said to be (right) *censored*. Then, the probability for this category should

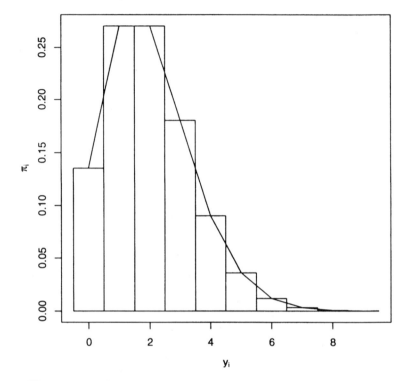

Fig. 4.5. The histogram and polygon for a Poisson distribution with $\mu = 2$.

be calculated as one minus the sum of the probabilities for all of the smaller counts.

This distribution has some important characteristics.

- The theoretical variance is equal to the mean, $\sigma_T^2 = \mu$.
- If each Y_i of a set of counts has a Poisson distribution with mean μ_i, then, the sum of the counts $Y_\bullet$, will also have a Poisson distribution with a mean that is the sum of the individual means: $\mu_\bullet$.

Example

In 1974, I conducted a study of factors affecting attendance and success at primary school in Bombay, India. A total of 73 randomly chosen schools was involved. Questionnaires were administered, in the five mother tongues used in these schools, to the 5920 children selected in second and third years (Standards). In the first seven columns of Table 4.3, some results from this study are displayed as a two-way contingency table classifying children in the third Standard by their age and social class. The corresponding histograms are plotted in Figure 4.6.

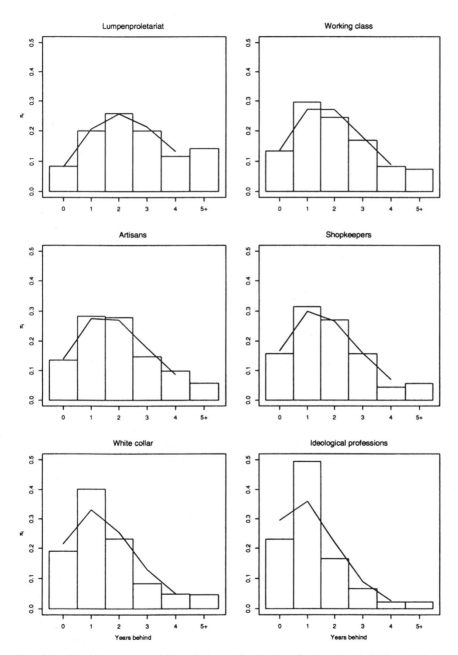

Fig. 4.6. The histograms and fitted Poisson distributions for the school children in Bombay in Table 4.3.

Table 4.3. Age of children in Standard 3 of primary school in Bombay in 1973, classified by social class. (Lindsey, 1978, p. 80)

Social class	Age 7	8	9	10	11	12+	$\bar{y}_\bullet$	s^2	AIC
Lumpen-proletariat	13	31	40	31	18	22	9.56	2.63	2.0
Working class	98	217	179	124	61	54	9.03	2.23	6.1
Artisans	33	69	68	36	24	14	8.99	2.07	2.0
Shopkeepers	25	50	43	25	7	9	8.87	1.94	2.9
White collar	87	184	106	38	22	21	8.56	1.80	15.4
Ideological professions	42	90	30	12	4	4	8.23	1.25	10.0

Table 4.4. Estimated mean and variance of the number of years children in Standard 3 of primary school in Bombay in 1973 are behind the norm, classified by social class, for the data of Table 4.3.

Social class	$\bar{y}_\bullet$	s^2
Lumpenproletariat	2.56	2.63
Working class	2.03	2.23
Artisans	1.99	2.07
Shopkeepers	1.87	1.94
White collar	1.56	1.80
Ideological professions	1.23	1.25

As we saw in Section 2.3.1, we could apply a polytomous logistic model to these data to see how age depends on social class. This model would supply useful information. However, it would have a large number of parameters (30, five for the multinomial distribution in each social class). Here, the response variable is quantitative, so that it is possible to explore what density function might describe it.

The estimated means and variances (from the original data, not those with the older children grouped, as in the table) are shown in the third and second last columns of Table 4.3. They both decrease down the column.

One additional piece of information is useful here. The age at which a child is supposed to enter primary school in Bombay is five. In Standard 3, they should be seven. Now, we can transform the response variable, the age, to the number of years behind this norm, by subtracting seven from all ages. In this way, we obtain a new variable, that is, a count, of years behind. The estimated means are transformed in the same way, but the estimated variances do not change. (Why?) These are both given in Table 4.4.

If we now compare these new estimated means with the variances, we see that they are very similar. This indicates that a Poisson distribution might be suitable. The curves for this distribution are also plotted in Figure 4.6. (Recall that, when fitting this distribution, the probability for 5+ will be one minus the sum of the probabilities for 0, 1, 2, 3, and 4.) Each of these models has only one parameter instead of five for the multinomial, so that the total number of estimated parameters for the table is six, instead of 30 if a polytomous logistic model were fitted.

The negative log normed likelihoods can be calculated using Equation (4.1). Then, the corresponding AICs for each social class are given in the last column of Table 4.3, each with one parameter. In each case, the saturated multinomial model has an AIC of 5 so that there are 4 d.f. Thus, although there is no evidence against a Poisson distribution for the lumpenproletariat, artisans, and shopkeepers, there is for the other three classes. This is visible in the graphs.   □

Over- and underdispersion   The Poisson distribution is especially interesting because it describes independent events happening *at random* to an individual. Fitting this distribution provides a standard which can allow one to detect departures from random independent events. One possibility is that some individuals are having too many events and others too few. Then, there will be too much variability in the responses for them to be independent and random. The second possibility is that all individuals are too similar in the numbers of events they have. Here, there will be too little variability.

When there is too much variability, one explanation could be that the probability of the event is not constant across individuals. This is known as *clustering* or *proneness*. (We saw, in Section 1.4.2, that clustering can also arise from the way in which the sample is chosen.) Another possible explanation is that the probability of an event changes over the time in which the events were occurring for each individual. This is known as *contagion*.

It is rarer to find that the variability is too small and that the events are distributed too uniformly across the individuals for them to be random. This is called *repulsion*, usually arising from lack of independence. The presence of a few events impedes further events from occurring.

An indication of lack of independence and randomness can be obtained by comparing the estimated and theoretical variances. For the Poisson distribution, we have seen that the theoretical variance equals the mean. Thus, a useful indicator for this is the ratio of the estimated variance to the estimated mean: $s^2/\hat{y}_\bullet$. This is called the *coefficient of dispersion*. Because the mean and variance should theoretically be equal for the Poisson distribution, this ratio should be about one. However, other distributions could also have a ratio of unity, so this would not demonstrate that the distribution is Poisson. On the other hand, departures from unity will indicate that the distribution is not likely to be Poisson, and in what way it is not.

If the coefficient of dispersion is considerably greater than one, say at least twice, there is too much variability in the counts of the events for them to be

Table 4.5. Residuals for the Poisson distributions fitted to the data for the six social classes in Table 4.3.

Age	Social class					
	LP	WC	A	SK	WC	IP
7	0.042	−0.188	−0.216	−0.319	−1.176	−1.602
8	−0.176	1.270	0.212	0.348	2.643	3.019
9	0.026	−1.377	0.244	0.075	−0.951	−1.579
10	−0.360	−0.681	−1.096	−0.062	−2.785	−1.056
11	−0.570	−0.578	0.607	−1.280	−0.173	−0.430
12+	1.306	2.567	0.562	1.420	3.846	2.033

random. They are said to be *overdispersed*. When it is considerably smaller, they are *underdispersed*.

Example (continued)
From Table 4.4, the coefficient of dispersion is close to one for all social classes. Thus, this does not help us here to determine why the Poisson distribution fits poorly for three of them. On the other hand, examination of the graphs in Figure 4.6 reveals the departures from the Poisson distribution for social classes where it does not fit. There are too many one year behind and too few two years behind. This is confirmed by inspection of the residuals in Table 4.5. Notice also that, in every social class, there are too many old children as compared with the Poisson distribution, although this is especially true for the same three classes where the model does not fit well.

If we provisionally accept that a Poisson distribution might be generating the counts of years behind for certain social classes in these data, we can see two different patterns. The children are, on average, getting closer to the normal age as we move down the table through the social classes. But, one social class near the top and the two at the bottom seem to have these events not being generated at random. Two separate mechanisms appear to be operating.

In the context of Bombay in 1974, we should not consider children to be behind in school because they failed one or more years. Instead, it will most often be the case that they have not attended school during those years, either by starting older than the norm, or by dropping out one or more times. Note also that many young children in this society work to provide family income.

We may now ask what distinguishes the three social classes where the Poisson distribution is least plausible. Parents in all these classes, as distinct from the other three, have jobs with steady income. However, in those classes where the Poisson model is acceptable, economic problems may arise at random times. Then, the help of the children will become necessary and they will not be able to attend school. This could explain the form of the distribution for these classes. On the other hand, the two 'middle' classes, the artisans and shopkeepers, often value education more than the working class, as a means of social promotion.

They make extra efforts to have their children attend school, when possible, which would explain their lower average age. □

When over- or underdispersion is present, more complex models than the Poisson distribution will be required (see Section 4.3.4).

Point processes If events are happening to individuals, then, these will usually not be occurring simultaneously, but as a series over time. This is called a *point process*, because events are happening only at certain points in time.

Now consider one simple special case. Suppose that the probability of an event remains constant over time, not depending on what happened before, and that the probability of several events occurring simultaneously (in a tight cluster) is negligible. Then, we have a *Poisson process*. Suppose that the unit of measurement of time is Δt; in fact, above I assumed this unit to be one. Then, Equation (4.2) can be rewritten as

$$\pi_i = f(y_i; \mu)$$
$$= \frac{e^{-\mu \Delta t} (\mu \Delta t)^{y_i}}{y_i!}$$

Then, μ is the mean number of events per unit time, called the *rate* or *intensity* of the events in the process. Thus, in this Poisson process model, the rate or intensity is assumed to remain constant over all time.

Example (continued)
The possible explanations given above for the Bombay children are only hypotheses. These would need to be studied more closely by obtaining data on the actual order of the years in which each child did not attend school and on the reasons why. For each child, there would be a point process, beginning at age five, where an event would be not attending school each successive year. One important question is whether this point process is a Poisson process for the lumpenproletariat, artisan, and shopkeeper classes. □

Relationship to the multinomial distribution An interesting relationship exists between the Poisson and multinomial distributions, although this is primarily useful for numerical calculations. Suppose that we have a number of independent counts y_i, each with a different mean μ_i, and that we condition on their total $y_\bullet$. We know that the latter will also have a Poisson distribution, with mean, $\mu_\bullet$. Then,

$$\Pr(y_1,\ldots,y_I|y_\bullet;\mu_1,\ldots,\mu_I) = \frac{\Pr(y_1,\ldots,y_I;\mu_1,\ldots,\mu_I)}{\Pr(y_\bullet;\mu_\bullet)}$$

$$= \frac{\Pi_i \frac{e^{-\mu_i}\mu_i^{y_i}}{y_i!}}{\frac{e^{-\mu_\bullet}\mu_\bullet^{y_\bullet}}{y_\bullet!}}$$

$$= \binom{y_\bullet}{y_1,\ldots,y_I} \frac{\Pi_i \mu_i^{y_i}}{\mu_\bullet^{y_\bullet}}$$

$$= \binom{y_\bullet}{y_1,\ldots,y_I} \prod_i \pi_i^{y_i}$$

where $\pi_i = \mu_i/\mu_\bullet$. Thus, any multinomial distribution, including all those used for the logistic and log linear models of Chapter 2, can also be modelled as Poisson distributions, if the total of the counts is held fixed, which is the case in such models anyway.

This also means that the theoretical mean and variance of a count in such a frequency table are both equal to $\mu_i = n_\bullet\pi_i$. For the Poisson distribution, the estimate of the theoretical standard deviation is $\sqrt{n_\bullet\widehat{\pi}_i}$. This is used in the denominator of the standardised residuals for models for these tables. Notice, however, that this is not the same as the estimate of the theoretical standard deviation for the multinomial distribution (Section 4.1.1), which is $\sqrt{n_\bullet\widehat{\pi}_i(1-\widehat{\pi}_i)}$.

4.3.2 GEOMETRIC DISTRIBUTION

With the Poisson distribution, we have seen how a count of the number of events occurring to an individual is usually an aggregation of events that actually occur as a process over time. In certain situations, for such a point process, the time y_i until the first event may be of particular interest. The number of time units before this first event is sometimes called the *waiting time*.

Under certain conditions, the study of such a duration can be framed in terms of counts. This occurs when the duration until the first event can be measured in some discrete equal-sized units of time. Then, in fact, we are counting the number of such time units without an event, stopping when the event first occurs. This will be a first example of modelling a duration that I shall look at in considerable detail in Section 4.5.

Suppose that the probability ν_1 of the event in a time unit, that is, its rate or intensity, remains constant over the period of observation. During y_i units of time, the event does not occur, each with probability $\nu_2 = 1 - \nu_1$ so that these are multiplied together. Finally, in the last time unit, the event occurs, with probability ν_1. This yields the geometric distribution

$$\pi_i = f(y_i;\nu_1)$$
$$= (1-\nu_1)^{y_i}\nu_1, \qquad 0\le\nu_1\le 1, \qquad 0\le y_i = i-1$$

The construction of this model is similar to that for the binomial and multinomial distributions in Section 1.3.4. However, here only one order is possible so that

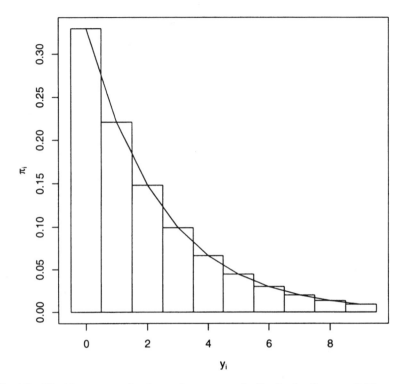

Fig. 4.7. The histogram and polygon for a geometric distribution for $\nu_1 = 0.33$ or $\mu = 2$.

the combinatorial is not necessary. An example of this distribution is plotted in Figure 4.7.

The theoretical mean and variance of the response y_i are, respectively,

$$\mu_T = \frac{1 - \nu_1}{\nu_1}$$

$$\sigma_T^2 = \frac{1 - \nu_1}{\nu_1^2}$$

Thus, the distribution can also be written in terms of the mean duration as

$$\pi_i = \frac{\mu^{y_i}}{(1 + \mu)^{1 + y_i}}, \qquad \mu > 0$$

where

$$\mu = \frac{1 - \nu_1}{\nu_1}$$

is the mean time until the first event.

This distribution describes the waiting time until the first event, counted in discrete equal periods of no event. It has the particular property that the probability of still waiting a time y_i for the event does not depend on the time, say y_0, already waited. The process has no memory:

$$\Pr(y_i + y_0 | y_0) = \Pr(y_i)$$

This is called the *Markov property*.

The maximum likelihood estimate of the geometric probability parameter is

$$\widehat{\nu}_1 = \frac{n_\bullet}{n_\bullet + \sum_i n_i y_i}$$

where the n_i are the frequencies of the counts y_i or, equivalently, that of the mean is $\hat{\mu} = \bar{y}_\bullet$.

Example
Table 4.6 gives the employment times of people aged between 25 and 44 recruited to the British Post Office in the first quarter of 1973. Departures were due to causes such as resignations, deaths, and so on. Let us assume that departure occurred in the given month so that y_i is one less than this. The histogram is plotted in Figure 4.8.

The mean duration is estimated to be $\hat{\mu} = 3.40$ months, or $\widehat{\nu}_1 = 0.227$. The fitted probabilities are shown in Table 4.6. The AIC is 36.5 with one parameter, as compared with 23 for the multinomial, indicating a poor fit. However, there are 22 d.f. Examination of the fitted curve in Figure 4.8, or the residuals in Table 4.6, indicates that there are too many people leaving after one, two, or three months and more than 15 months, whereas there are too few after between four and 14 months. Thus, the probability of departure does not appear to be constant over time.   □

With a bit of imagination, the geometric distribution can also be applied in situations not involving durations.

Example
In the effort to improve urban methods of transportation, studies have been made for quite some time in an attempt to estimate the number of passengers carried by each car in urban traffic. Table 4.7 gives the number of occupants, including the driver, in passenger cars passing the intersection of Wilshire and Bundy Boulevards in Los Angeles, USA, on Tuesday, 24 March 1959, between 10:00 and 10:40.

Let us reconstruct the response variable, the number of occupants in the car, taking the event to be the driver, always present. Then, the new response, the 'waiting time', is the number of passengers, that is, the number of occupants minus one. Thus, we are saying that, in a car i, we must 'wait' while observing y_i

Table 4.6. Duration of employment of recruits to the British Post Office, with the fitted geometric distribution and residuals. (Burridge, 1981)

Months	Employees	Multinomial	Geometric	Residual
1	52	0.217	0.227	−0.338
2	46	0.192	0.176	0.598
3	50	0.208	0.136	3.057
4	27	0.113	0.105	0.366
5	15	0.063	0.081	−1.009
6	12	0.050	0.063	−0.783
7	8	0.033	0.048	−1.062
8	4	0.017	0.037	−1.662
9	5	0.021	0.029	−0.737
10	1	0.004	0.022	−1.885
11	0	0.000	0.017	−2.037
12	2	0.008	0.013	−0.674
13	1	0.004	0.010	−0.939
14	0	0.000	0.008	−1.384
15	0	0.000	0.006	−1.217
16	2	0.008	0.005	0.800
17	2	0.008	0.004	1.186
18	1	0.004	0.003	0.383
19	5	0.021	0.002	6.151
20	1	0.004	0.002	0.926
21	4	0.017	0.001	6.557
22	1	0.004	0.001	1.530
23	1	0.004	0.001	1.868
24	0	0.000	0.001	−0.382

Table 4.7. The number of occupants in passenger cars observed at an intersection in Los Angeles during 40 minutes on the morning of 24 March 1959, with the fitted geometric distribution and the residuals. (Derman *et al.*, 1973, p. 278, from Haight)

Occupants	Cars	Multinomial	Geometric	Residual
1	678	0.671	0.658	0.487
2	227	0.225	0.225	−0.030
3	56	0.055	0.077	−2.466
4	28	0.028	0.026	0.277
5	8	0.008	0.009	−0.359
6+	14	0.014	0.005	4.275

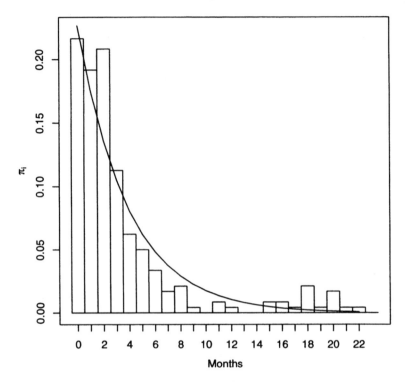

Fig. 4.8. The histogram and fitted geometric distribution for the post office employment data in Table 4.6.

passengers in order finally to observe the driver. The data are plotted in this form in Figure 4.9.

The maximum likelihood estimate is $\hat{\nu_1} = 0.658$ or $\hat{\mu} = 0.519$, the mean number of passengers in a car. The probabilities for the multinomial and geometric distributions are also given in Table 4.7 and plotted in Figure 4.9. The geometric distribution appears to follow the histogram fairly closely. This is not, however, confirmed by the AIC of 10.5 with one parameter, as compared with 5 for the multinomial, with 4 d.f. The residuals in Table 4.7 indicate that there are too few cars with two passengers and too many with five or more.   □

4.3.3   BINOMIAL DISTRIBUTION

In the geometric distribution, two types of events in fact occur (with probabilities ν_1 and $\nu_2 = 1 - \nu_1$), although only one is of interest; we stop observing when it occurs. On the other hand, the total number of events of the kind not of interest is not fixed. Now let us look at another model for data where two types of events

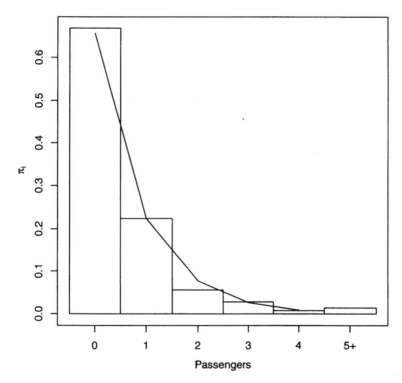

Fig. 4.9. The histogram and fitted geometric distribution for the car passenger data in Table 4.7.

are counted, but where the total number of events will be fixed. All events are assumed to be independent of each other. As usual, this is a strong assumption which often may not be justified. Nevertheless, it is often useful to fit such a model in order to investigate how the assumptions are violated.

Examples
   Students are given a test with a fixed number m of questions. Each student i will have a certain number y_i of correct answers and the remaining $m - y_i$ wrong.
   A sample of families, each with m children, is studied. A family i will have y_i boys and $m - y_i$ girls.
   The assumption of independence among events may be more plausible in the second case than the first. (Why?)   □

   Under the above assumptions of two independent types of events, the binomial distribution provides the simplest model. I introduced this distribution in Section 1.3.4 and also repeated some relevant results for the multinomial distribution in

Section 4.1.1. When there is only one observation from a binomial distribution, it is often called a Bernoulli trial.

Until now, especially in Chapter 2, I have used the binomial distribution to describe the *frequencies* of responses of a binary categorical variable. It will only be appropriate to use it to model *counts* if the events are *independent* events for an individual. Thus, the binomial distribution will provide a means of checking this independence assumption.

Suppose that any individual i has m events of which y_i are of the first type and the rest of the second type. This response y_i for an individual can take the values $0, 1, \ldots, m$. Thus, the number of different response categories will be $I = m + 1$: there will be one more multinomial category than the total number of events. The multinomial probabilities, under this binomial model, can be estimated using

$$\pi_i = f(y_i; \nu_1) \qquad 0 \leq \nu_1 \leq 1$$
$$= \binom{m}{y_i} \nu_1^{y_i} (1 - \nu_1)^{m - y_i} \qquad 0 \leq y_i = i - 1 \leq m$$

from Equation (1.9). However here, as with the geometric distribution, I represent the Bernoulli probability (the binomial parameter) by ν_1 instead of π_1 to avoid confusion with the multinomial probability π_i.

As we have seen in Section 1.3.4, the theoretical mean number of events of the first type and its theoretical variance are, respectively,

$$\mu_T = m\nu_1$$
$$\sigma_T^2 = m\nu_1(1 - \nu_1)$$

In the present context, the maximum likelihood estimate of the binomial parameter is

$$\widehat{\nu_1} = \frac{\sum_i n_i y_i}{mn_\bullet}$$

where n_i are the frequencies of the counts y_i. The numerator is the total number of the first type of event and the denominator the total number of all events.

Example

Study of the human sex ratio is greatly indebted to the data collected by Geissler (1889) on the distributions of the sexes of children in families in Saxony in 1876–1885. The data contain the size of family, with the sex of all children, at the time of registration of the birth of a child. They, thus, do not necessarily refer to complete families because the parents could have more children in the future.

Here, I shall consider the subset of the data concerning the numbers of male and female children in families of a fixed size 12, so that $I = 13$. The data are given in Table 4.8. The first column contains counts and the second their frequencies.

Table 4.8. Counts of male children in 6115 families of size 12 in Saxony, with the fitted binomial and beta-binomial distributions and their standardised residuals. (Sokal and Rohlf, 1969, p. 80)

Males	Families	Multin.	Binom.	Residual	Beta-binom.	Residual
0	3	0.000	0.000	2.140	0.000	0.426
1	24	0.004	0.002	3.423	0.004	0.302
2	104	0.017	0.012	3.793	0.017	−0.078
3	286	0.047	0.042	1.708	0.051	−1.409
4	670	0.110	0.103	1.676	0.107	0.558
5	1033	0.169	0.177	−1.591	0.169	−0.102
6	1343	0.220	0.224	−0.658	0.206	2.395
7	1112	0.182	0.207	−4.323	0.193	−2.042
8	829	0.136	0.140	−0.864	0.140	−0.840
9	478	0.078	0.067	3.361	0.076	0.751
10	181	0.030	0.023	4.181	0.029	0.237
11	45	0.007	0.004	3.706	0.007	0.187
12	7	0.001	0.000	3.038	0.001	0.785

If we fit a binomial distribution, we make two assumptions in this model: that the probability of a child of a given sex, for successive births, is independent, that is, constant, within each family and that it is the same for all families.

From these data, we can calculate the probability of a child of given sex, which may not be one half! Thus, the estimated probability of a male is $\hat{\nu}_1 = 0.519$. The relative frequencies, $\hat{\pi}_i$, which are the maximum likelihood estimates for the multinomial distribution, and the corresponding estimates, $\tilde{\pi}_i$, when the binomial model is applied, are also given in Table 4.8. The estimated binomial model is plotted as a density function on the histogram in Figure 4.10.

The negative log normed likelihood for this model is calculated to be 48.5 with $(I - 1) - 1 = 11$ d.f. The AIC is 49.5 as compared with 12 for the saturated model, giving strong evidence that the binomial distribution does not fit well. (Here is a good example where more decimals are required for the π_i than those given in the table in order to obtain an accurate value for the log likelihoods.)

The standardised residuals are shown in the fifth column of Table 4.8. We see that the frequencies in the two tails of the distribution are underestimated by the binomial model and those in the centre overestimated. There are more families with children of predominantly one sex than would be expected if births were independent with constant probability.   □

Overdispersion As we have seen for the Poisson distribution, when an observed distribution of counts has thicker tails than expected according to a standard model, overdispersion is said to be present. As with the Poisson distribution, when the standard is the binomial distribution, overdispersion may be due to clus-

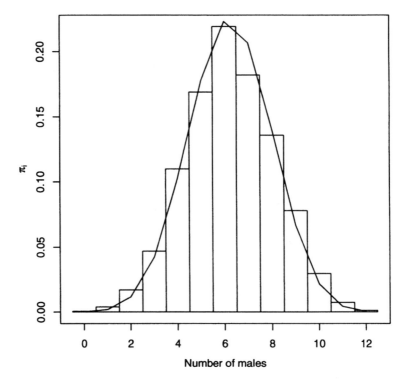

Fig. 4.10. The histogram and binomial density function for the number of male children in families of 12.

tering or proneness. Again, an indication can be obtained from the ratio of the estimated to the theoretical variance. Such systematic departures from the model should be taken into account in the development of a more complex model (see Section 4.3.5).

Example (continued)
For these data, the mean number of males is estimated as $\bar{y}_\bullet = 6.23$ and the variance to be $s^2 = 3.49$, whereas the estimated theoretical binomial variance is $12\hat{\nu}_1(1 - \hat{\nu}_1) = 3.00$. The estimated variance is somewhat larger than the theoretical one, although not really large enough to indicate overdispersion. A better indicator of lack of fit is the comparison of AICs given above.

The thick tails could arise from one of at least two different data generating mechanisms, linked to our two assumptions stated above. The probability of having a male child may vary across the different families, with some families having higher probabilities of one of the sexes or the other. Or the probabilities may be the same to start with in all families, but change if a series of children of one sex happens to be born, although this does not seem to be very biologically plausible. Still a third possibility would be that some families of size 12 might result precisely because the parents already had mainly children of one sex and wanted a more balanced family. They continued to have children, although they could not influence the sex of each one. In order to study these mechanisms, and to distinguish between these models, the order of the births by sex in each family would need to be available and some sort of point process model constructed.   □

4.3.4 NEGATIVE BINOMIAL DISTRIBUTION

We saw above that the geometric distribution describes the waiting time until the first event. This can be generalised to obtain the waiting time until say c events have occurred. The result is called the negative binomial distribution. The response variable y_i is the number of time units without an event before all c events have occurred so that the total time is $y_i + c$. The probability density function is

$$\pi_i = f(y_i; \nu_1) \qquad 0 \leq \nu_1 \leq 1$$
$$= \binom{c + y_i - 1}{y_i} \nu_1^c (1 - \nu_1)^{y_i} \qquad 0 \leq y_i = i - 1 \qquad (4.3)$$

where ν_1 is the Bernoulli probability of the event. Thus, the geometric distribution is the special case when $c = 1$.

Sampling distribution   Another interpretation for the negative binomial distribution is in terms of Bernoulli events. It gives the distribution of the number of events of the second kind, with probability $\nu_2 = 1 - \nu_1$, before a fixed number c events of the first kind occur. Thus, Equation (4.3) resembles a binomial distribution. However, it is not the same because the number of one of the events is fixed instead of the total number of both kinds of events.

Example
In a standard sample survey, we fix the total number of observations, the sample size. Then, if the responses recorded are binary and independent, a binomial distribution may be appropriate. Now instead of a fixed total number of observations, suppose that we change our sampling design and fix the number c of positive responses to the question that we require in our sample. Then, the number of negative responses, before obtaining c positive ones, may follow a negative binomial distribution.   □

Like the Poisson distribution, and in contrast to the binomial distribution, here the response variable can take any non-negative integer value. An example is plotted in Figure 4.11. (This may also be compared with Figures 4.5 and 4.7. The latter is a negative binomial distribution with $c = 1$ and $\mu = 2$. We see that, as c decreases, the probabilities of small counts increase.)

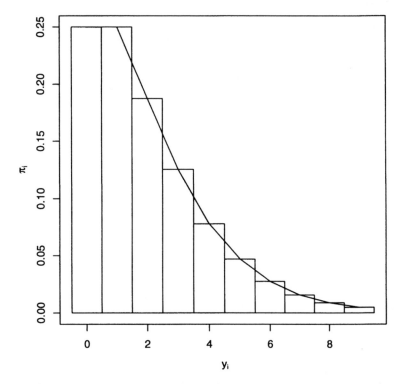

Fig. 4.11. Negative binomial distribution for $c = 2$ and $\nu_1 = 0.5$ or $\mu = 2$.

The maximum likelihood estimate of the negative binomial parameter is

$$\widehat{\nu_1} = \frac{n_\bullet c}{n_\bullet c + \sum_i n_i y_i} \tag{4.4}$$

As with the geometric and binomial distributions, the numerator is the total number of the first type of event and the denominator the total number of all events.

The theoretical mean and variance of the number of counts of the second type, the response variable, are, respectively,

$$\mu_T = \frac{c(1 - \nu_1)}{\nu_1}$$

$$\sigma_T^2 = \frac{c(1 - \nu_1)}{\nu_1^2}$$

Application to overdispersion   This distribution can be generalised by allowing c to become an unknown parameter, allowed to take nonintegral values. Because it is now a parameter, I shall call it γ. We are now only counting one type of event, those which were the second type above. Equation (4.3) can be rewritten as

$$\pi_i = f(y_i; \nu_1, \gamma) \qquad 0 \le \nu_1 \le 1 \qquad \gamma > 0$$

$$= \frac{\Gamma(\gamma + y_i)}{y_i! \Gamma(\gamma)} \nu_1^\gamma (1 - \nu_1)^{y_i} \qquad 0 \le y_i = i - 1$$

$$= \frac{\Gamma(\gamma + y_i)}{y_i! \Gamma(\gamma)} \frac{\gamma^\gamma \mu^{y_i}}{(\gamma + \mu)^{\gamma + y_i}} \qquad \mu > 0$$

where

$$\mu = \frac{\gamma(1 - \nu_1)}{\nu_1}$$

the theoretical mean number of events given above. Then, the theoretical variance in this parametrisation will be

$$\sigma_T^2 = \mu + \frac{\mu^2}{\gamma}$$

Thus, the variance will always be larger than that for the Poisson distribution, which is μ.

In this form of the distribution, $\Gamma(\cdot)$ is the gamma function, a generalisation of a factorial to nonintegral values, but still retaining a recursive relationship:

$$\Gamma(a + 1) = a\Gamma(a)$$

Thus, if a is an integer,

$$\Gamma(a + 1) = a!$$

Useful values are given in the Appendix, in Table A.5. This function will appear in a number of distributions in this chapter.

The negative binomial distribution is commonly used in place of the Poisson distribution for count data of an event when there is overdispersion, perhaps due to proneness or clustering. In other words, it applies when events counted are not independent. This has several possible interpretations that cannot be distinguished without having available the timings of the events for each individual. The most common is that the events of each individual follow a Poisson distribution but that the mean numbers of events vary among the individuals. Thus, individuals prone to have the event have larger means (although we do not know who they are). Then, the dependence or *correlation* (Section 5.4) among events on each individual is $\rho = 1/\gamma$.

In the context of overdispersion, Figure 4.11 can be compared with that for the Poisson distribution in Figure 4.5; both have a mean of two, but the negative binomial is less sharply peaked, more dispersed. For a given mean μ, the negative

Table 4.9. Numbers of units purchased in 4 weeks of a particular brand of item, with probability estimates and residuals for two models. (Derman *et al.*, 1973, p. 291, from Chatfield)

Units	Consumers	Multin.	Poisson	Residual	Neg. Bin.	Residual
0	1671	0.955	0.915	1.732	0.955	0.001
1	43	0.025	0.081	−8.300	0.026	−0.335
2	19	0.011	0.004	5.074	0.009	0.663
3	9	0.005	0.000	20.467	0.004	0.468
4	2	0.001	0.000	31.144	0.002	−1.023
5	3	0.002	0.000	351.707	0.001	0.478
6	1	0.001	0.000	964.937	0.001	−0.285
7	0	0.000	0.000	−0.000	0.000	−0.892
8	0	0.000	0.000	−0.000	0.000	−0.697
9	2	0.001	0.000	1643630.	0.000	3.095
10+	0	0.000	0.000	-0.000	0.000	−0.435

binomial can have many forms as γ varies, whereas the Poisson can have only one.

Unfortunately, the maximum likelihood estimates simultaneously for ν_1 or μ and γ are rather difficult to obtain without a computer. An approximation to them is

$$\hat{\nu}_1 \doteq \frac{\bar{y}_\bullet}{s^2}$$

$$\hat{\mu} \doteq \bar{y}_\bullet$$

$$\hat{\gamma} \doteq \frac{\bar{y}_\bullet^2}{s^2 - \bar{y}_\bullet}$$

These are called moment estimates of the parameters. They can give negative estimates of γ and estimates of $\nu_1 > 1$, but only if there is underdispersion, $\bar{y}_\bullet > s^2$. As well, they are not maximum likelihood estimates, so that they do not maximise the likelihood or minimise the AIC.

Example

A random sample of consumers was asked to keep a record of their own purchases over a four-week period. Table 4.9 gives numbers of units of a particular package size of a particular brand of one item purchased within that time.

The estimated mean and variance are $\bar{y}_\bullet = 0.0886$ and $s^2 = 0.281$. The ratio, $0.281/0.0886 = 3.17$, the coefficient of dispersion, indicates a large amount of overdispersion. This tells us that there is too much variability in the number of purchases by different consumers for a Poisson distribution of random purchasing to be applicable. It is quite possible that the probability of purchase varies among the participants, especially because so many made no purchases (Figure 4.12).

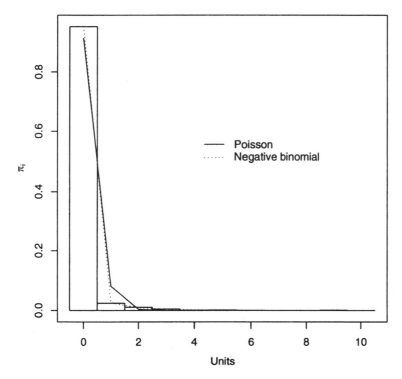

Fig. 4.12. The histogram and Poisson and geometric density functions for the purchases in Table 4.9.

For the negative binomial distribution, the approximate parameter estimates are $\widehat{\nu}_1 \doteq 0.316$ and $\widehat{\gamma} \doteq 0.041$. The maximum likelihood estimates are very close: $\widehat{\nu}_1 = 0.306$ and $\widehat{\gamma} = 0.039$. The estimated probabilities for the multinomial, Poisson, and negative binomial distributions are also given in Table 4.9. We see that the negative binomial estimates are much closer to the relative frequencies than are the Poisson. There are too many consumers not buying any units or buying several, and not enough buying one unit for a Poisson distribution with an average of 0.0886 units per consumer.

The AIC for the Poisson distribution is 190.3 with one parameter, whereas it is 7.1 for the negative binomial with two parameters, as compared with an AIC of 10 for the multinomial. This confirms that, with one more estimated parameter, the latter fits much better than the former (using up one of the 9 d.f. of the Poisson distribution). Note, however, that there are several categories with small or zero frequencies, which should be grouped if a significance test is to be applied. Of course, the larger frequencies must not be grouped, because that could distort the form of the distribution. □

4.3.5   BETA-BINOMIAL DISTRIBUTION

As we have seen, the negative binomial distribution can be used when data are
overdispersed with respect to the Poisson distribution. One interpretation was
that the mean parameter of the Poisson distribution varies among individuals. We
can also make a similar assumption about the binomial distribution: the probabilities of the two types of events vary among individuals. This will also result
in overdispersed data, with respect to the binomial distribution. Then, the corresponding distribution is the beta-binomial distribution:

$$\pi_i = f(y_i; \nu_1, \gamma), \qquad \gamma > 0$$
$$= \binom{m}{y_i} \frac{B[y_i + \nu_1\gamma, m - y_i + (1 - \nu_1)\gamma]}{B[\nu_1\gamma, (1 - \nu_1)\gamma]}$$

where ν_1 is the usual Bernoulli probability for the binary events (although here
they are not independent), whereas $\rho = 1/(1+\gamma)$ is the dependence or *correlation*
(Section 5.4) among these events. Here,

$$B(a,b) = \frac{\Gamma(a)\Gamma(b)}{\Gamma(a+b)}$$

is called the beta function. Then, the theoretical mean is the same as for the
binomial distribution ($\mu_T = m\nu_1$), whereas the theoretical variance is

$$\sigma_T^2 = m\nu_1(1 - \nu_1)[1 + (m - 1)\rho]$$

This must be larger than that for the binomial distribution, $m\nu_1(1 - \nu_1)$, because
$\rho > 0$.

As with the negative binomial distribution, the interpretation of this model
given above is the common, but not the only, one. The time series of events for
each individual, rather than simply the aggregated counts, is necessary to distinguish among the possibilities.

Approximate estimates of the parameters can be obtained by using

$$\hat{\nu_1} \doteq \frac{\sum_i n_i y_i}{mn_\bullet}$$
$$\hat{\gamma} \doteq \frac{m^2\hat{\nu_1}(1 - \hat{\nu_1}) - s^2}{s^2 - m\hat{\nu_1}(1 - \hat{\nu_1})}$$
$$\hat{\rho} \doteq \frac{s^2 - m\hat{\nu_1}(1 - \hat{\nu_1})}{m(m - 1)\hat{\nu_1}(1 - \hat{\nu_1})}$$

Because ρ must be non-negative, the empirical variance must be larger than the
estimated binomial variance for this to work.

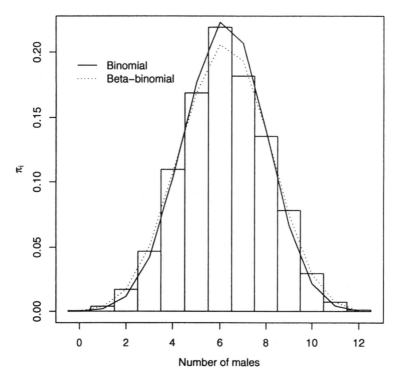

Fig. 4.13. The histogram and binomial and beta-binomial density functions for the number of male children in families of 12.

Example (continued)
We saw in Section 4.3.3 that the data on the sex of children were overdispersed with respect to the binomial distribution. Hence, it will be informative to try the beta-binomial distribution, using one of the 11 d.f. of the binomial. Here, the approximate and maximum likelihood estimates are virtually identical: $\widehat{\nu}_1 = 0.519$ and $\hat{\rho} = 0.015$. The fitted probabilities and residuals were given in Table 4.8. The former are plotted in Figure 4.13. The thicker tails of the beta-binomial distribution, as compared with the binomial, are visible.

The AIC for the beta-binomial distribution, with two parameters, is 9.2, as compared with 49.5 for the binomial with one parameter and 12 for the multinomial. Thus, the model fits well. However, the number of families with equal numbers of boys and girls is underestimated with this model.   □

4.4   Distributions for measurement errors

In the scientific measurement of inert substances, for example in physics or chemistry, most of the variability is produced by the measurement procedure, the substances involved usually being very homogeneous. These *measurement errors* will usually vary symmetrically about the 'true' value. In contrast, in the study of living beings, the main sources of variability arise from the differences among individuals. Thus, although measurement errors are present in any observations, in many situations they are drowned by the other, much larger, sources of variability. In this section, I shall look at probability distributions suitable for the special situation where measurement error is the predominant type of variability.

4.4.1   NORMAL DISTRIBUTION

The best known distribution for measurement error is the normal or Gaussian. This distribution is primarily important for its nice mathematical properties. It is much over used in many areas of research for this very reason. In many scientific fields, few response variables are available that might follow such a distribution; count (Section 4.3) and duration (Section 4.5) responses are more common.

Like all symmetric distributions to be presented here, the normal distribution describes a continuous response variable, taking any real value, positive or negative. A response variable that is the result of a large number of small accumulating, unknown, additive influences could follow this distribution.

Example
Human heights or weights are determined by many genetic, nutritional, environmental, and social factors. However, only heights generally have a symmetric distribution. (Why?) Nevertheless, even these response variables do not provide realistic examples for such a distribution, because heights and weights cannot be negative.   □

The normal distribution has the well-known 'bell-shaped' form. In theory, the response variable is continuous, with no restriction on its values. However, as with all continuous variables, in practice it can only be observed up to the precision Δ_i of the measuring instrument involved. Thus, recording an observation y_i means, in fact, that it lies in the interval $(y_i - \Delta_i/2, y_i + \Delta_i/2)$ for some known Δ_i determined by the instrument used.

The normal distribution,

$$\pi_i = f(y_i; \mu, \sigma^2)\Delta_i$$

$$= \frac{1}{\sqrt{2\pi\sigma^2}} e^{-\frac{1}{2}\left(\frac{y_i-\mu}{\sigma}\right)^2} \Delta_i \tag{4.5}$$

is often written $y_i \sim N(\mu, \sigma^2)$. Note that, here, π without a subscript is not a parameter, but the constant, $3.14159\cdots$. The theoretical mean μ and variance σ^2

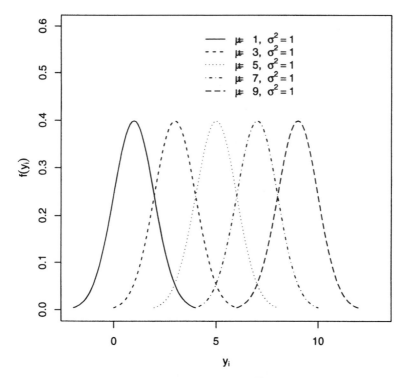

Fig. 4.14. Normal density functions with $\sigma^2 = 1$ and various means.

are both explicit parameters. The first is a location parameter determining the position of the distribution and the second is a dispersion parameter determining its width (the variability). The square root of the variance, σ, is called the standard deviation (Section 1.2.2).

This distribution has a special property, characteristic of many symmetric distributions: the location and dispersion parameters can each be allowed to vary without affecting the other. The first only controls the position and the second only the shape. In contrast, for most common distributions, such as those for counts above, if the location changes, not only the position, but also the *shape* of the distribution must change. Here, that is not the case, as can be seen in Figure 4.14 when the mean is varied.

As a result, all normal distributions can be standardised to have the same location and the same variance, generally set, respectively, to $\mu = 0$ and $\sigma^2 = 1$. This is accomplished by transforming the response variable to a *standard normal deviate*:

$$z_i = \frac{y_i - \mu}{\sigma}$$

Here, μ locates the response and σ scales it. Notice that y_i only appears in this form in the density. This is characteristic of many symmetric distributions; they are members of the location–scale family (Section 4.7.1).

Substituting the standard normal deviate into Equation (4.5) with $\sigma^2 = 1$, we obtain the standard normal density function,

$$f(z_i; 0, 1) = \frac{1}{\sqrt{2\pi}} e^{-\frac{z_i^2}{2}} \tag{4.6}$$

This has traditionally been called the error function.

Likelihood function   For $n_\bullet$ independent observations, the likelihood function is

$$L(\mu, \sigma^2) = \prod_i \left[\frac{1}{\sqrt{2\pi\sigma^2}} e^{-\frac{1}{2\sigma^2}(y_i - \mu)^2} \Delta_i \right]^{n_i}$$

$$= \frac{e^{-\frac{1}{2\sigma^2} \sum_i n_i (y_i - \mu)^2}}{(2\pi\sigma^2)^{\frac{n_\bullet}{2}}} \prod_i \Delta_i^{n_i}$$

For any fixed value of μ, the maximum of this likelihood occurs when $\sigma^2 = \sum_i n_i (y_i - \mu)^2 / n_\bullet$. Thus, we can rewrite the likelihood function only in terms of μ, a *profile* likelihood (Section 3.2.3):

$$L(\mu) = \frac{e^{-\frac{n_\bullet}{2}}}{[2\pi \sum_i n_i (y_i - \mu)^2 / n_\bullet]^{\frac{n_\bullet}{2}}} \prod_i \Delta_i^{n_i}$$

Thus, the normed profile likelihood for μ has a particularly simple form:

$$R(\mu) = \left[\frac{\sum_i n_i (y_i - \mu)^2}{\sum_i n_i (y_i - \hat{\mu})^2} \right]^{-\frac{n_\bullet}{2}}$$

Maximising either of these expressions with respect to μ is equivalent to minimising

$$\sum_i n_i (y_i - \mu)^2$$

(Why?) Its minimum is the usual maximum likelihood estimate, $\hat{\mu} = \bar{y}_\bullet$. This has often been called 'least squares' estimation.

Next, we can obtain the maximum likelihood estimate of the variance by substituting $\hat{\mu} = \bar{y}_\bullet$ into $\sigma^2 = \sum_i n_i (y_i - \mu)^2 / n_\bullet$. This gives $\hat{\sigma}^2 = s^2$. Sometimes the variance is estimated by the conditional (on $\bar{y}_\bullet$) maximum likelihood estimate,

$$\hat{\sigma}^2 = \frac{1}{n_\bullet - 1} \sum_i n_i (y_i - \bar{y}_\bullet)^2$$

$$= \frac{n_\bullet}{n_\bullet - 1} s^2$$

Table 4.10. Times (hours) reported spent doing a homework assignment by the first class taking this course in 1976, with the fitted normal distribution and the residuals.

Hours	Students	Multinomial	Normal	Residual
5–9	3	0.115	0.105	0.159
10–14	8	0.308	0.198	1.260
15–19	6	0.231	0.250	−0.199
20–24	5	0.192	0.213	−0.233
25–29	1	0.038	0.123	−1.225
30–34	2	0.077	0.047	0.691
35–39	0	0.000	0.012	−0.567
40–44	1	0.038	0.002	3.974

Then, the negative log normed likelihood is

$$-\log[R(\mu)] = \frac{n_{\bullet}}{2} \log\left(\frac{\tilde{\sigma}^2}{\hat{\sigma}^2}\right) \tag{4.7}$$

and the deviance is twice this, where $\tilde{\sigma}^2 = \sum_i n_i (y_i - \mu)^2 / n_{\bullet}$ for some value of μ of interest.

Although the normal density function in Equation (4.5) appears complex, we see that the maximum likelihood estimates can easily be obtained through 'least squares'. This explains why this distribution is so widely used even though it is not scientifically appropriate in many applications.

Example
Table 4.10 gives the times spent doing a homework assignment by the first class taking this course in 1976, along with the relative frequencies and the estimated normal distribution. (Raw data are given in Table 5.2 below.) The parameter estimates, calculated from the ungrouped data, are $\hat{\mu} = 17.48$ hours and $\hat{\sigma}^2 = 63.28$. For the raw data, the unit of measurement is $\Delta_i = 0.5$ hours. However, for the grouped data in this table, we must use $\Delta_i = 5$ hours. The estimated density function for the normal distribution is plotted with the histogram in Figure 4.15. In fact, it continues to the left, giving positive probability to negative hours of homework! This is an excellent theoretical reason for not using the normal distribution for such data (see Section 4.5 below).

As might be expected, because these are duration times, the histogram is not symmetric. This is reflected in the standardised residuals given in the table, where small and very large times are underestimated. The AIC is 7.4, as compared with 7 for the multinomial, with 5 d.f. Thus, here there is little evidence for or against a normal distribution. (However, remember that there are only 26 observations grouped into eight categories and that these data are not a random sample from some population.) □

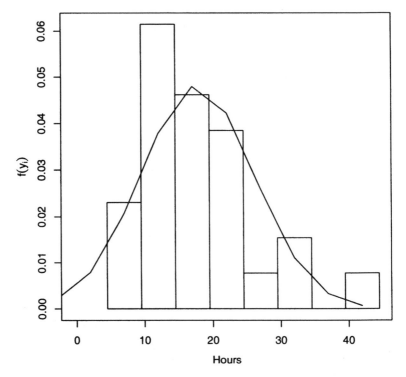

Fig. 4.15. The histogram and normal density function for the number of hours taken to do a homework assignment.

An approximate normed likelihood function, based on the standard error, was given in Equation (3.4). (As would be expected, the standard error of μ is given by $\sigma/\sqrt{n_\bullet}$.) There, z was a standard normal deviate and the approximate likelihood had the same form as that from a standard normal density of Equation (4.6). Thus, that approximation is exact only for the normal distribution with known variance. However, it is also very close to the normed profile likelihood for the mean given in Equation (4.7).

Cumulative probabilities In contrast to the density function, the cumulative distribution function (Section 1.3.4) for this distribution cannot be written in an explicit form. It involves a complex integral. Because of the common use of this distribution, tables of the integral are widely available and most statistical software can provide any required values. Let us look more closely at the way in which this can be used.

Because of the property that all normal distributions can be standardised to have the same mean 0 and variance 1, the probability that a response lies in any given range, for *any* normal distribution, can be obtained from a single table. These values are given in Table A.4 of the Appendix. Because this distribution is symmetric, $\Pr(-c < -z_i < 0) = \Pr(0 < z_i < c)$ so that only positive values of the standard normal deviate z_i need to be provided. Note that values of the cumulative distribution function $F(y)$ are not directly given in the table, but rather $0.5 - F(y)$.

Examples
Suppose that we have a normal distribution with mean 30 and variance 900. The standard normal deviate is

$$z_i = \frac{y_i - 30}{30}$$

Then, from Table A.4, we obtain

$$\Pr(0 < y_i < 30) = \Pr(-1 < z_i < 0)$$
$$= 0.341$$

by looking at the second column in the line for $|z_i| = 1.0$. If the two values of z_i are on opposite sides of zero, we add them, to obtain

$$\Pr(0 < y_i < 40) = \Pr(-1 < z_i < 0.33)$$
$$= 0.341 + 0.129$$
$$= 0.470$$

by looking, as well, at the line for $|z_i| = 0.3$ and moving across to the fifth column for 0.03, so that $|z_i| = 0.33$, whereas if they are on the same side of zero, we subtract,

$$\Pr(0 < y_i < 20) = \Pr(-1 < z_i < -0.33)$$
$$= 0.341 - 0.129$$
$$= 0.212$$

The total probability on one side of the mean, $z_i = 0$, is one half. If we want only the extreme part of this, we subtract, to obtain

$$\Pr(y_i < 0) = \Pr(z_i < -1)$$
$$= 0.5 - 0.341$$
$$= 0.159$$

□

Because of the mathematical ease of using the normal distribution, certain modifications of it, for asymmetric or skewed distributions, are available. I shall discuss some of them in Section 4.6 below.

Table 4.11. Twenty observations randomly generated from a Cauchy distribution with $\mu = 20$ and $\sigma = 2$.

17	12	18	63	25	16	21	17	18	19
19	19	17	16	20	22	25	20	38	21

Robustness   As we saw in Section 3.5, one important aspect of modelling building is to determine whether or not a given model adequately represents the available data. Certain distributions can produce much more extreme values than others. Thus, the thicker the tails, the more probable are values far from the centre of the distribution. Observations that appear to be extreme (outliers, see Section 3.5.2) for a thin-tailed distribution, such as the normal distribution, may be relatively probable in a more appropriate model based on a thicker tailed distribution. Such a model is said to be *robust* to extreme ('outlying') values.

Example
Consider the 20 observations shown in Table 4.11. At first glance, one might think that the fourth observation, and perhaps the second last, are outliers. Certainly, if a normal distribution were involved, they would be. However, they are quite probable if the model generating the data has thick tails.   □

I shall now present four symmetric distributions that have thicker tails than the normal distribution. They may be useful, for example, in situations where the substance being measured was contaminated in some particular way.

4.4.2   LOGISTIC DISTRIBUTION

A second symmetric distribution has a graphical form very similar to that of the normal distribution although the mathematical function is quite different. This is the logistic distribution with probability density function

$$f(y_i;\mu,\sigma') = \frac{e^{-\frac{y_i-\mu}{\sigma'}}}{\left[1+e^{-\frac{y_i-\mu}{\sigma'}}\right]^2}$$

However, because in this formulation $\sigma = \sigma'\pi/\sqrt{3}$ is the standard deviation, I shall instead use the more complex form

$$\pi_i = f(y_i;\mu,\sigma)\Delta_i$$
$$= \frac{\pi e^{-\frac{\pi}{\sqrt{3}}\left(\frac{y_i-\mu}{\sigma}\right)}\Delta_i}{\sqrt{3}\sigma\left[1+e^{-\frac{\pi}{\sqrt{3}}\left(\frac{y_i-\mu}{\sigma}\right)}\right]^2} \tag{4.8}$$

where, again, π is a constant. It is plotted in Figure 4.16, along with a normal distribution having the same mean and variance. The logistic distribution is slightly

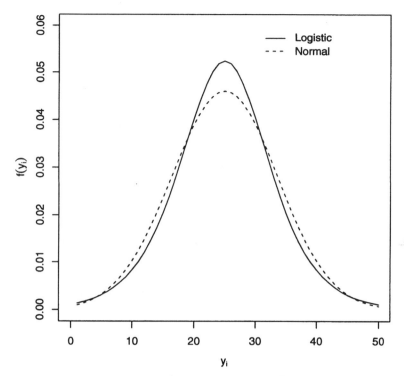

Fig. 4.16. The logistic density function with $\mu = 25$ and $\sigma^2 = 75$, and the corresponding normal density with the same parameter values.

narrower than the normal, and has thicker tails, although the theoretical mean and standard deviation are the same explicit parameters, μ and σ in both cases.

Although the two parameters are the theoretical mean and standard deviation, the empirical mean and standard deviation are not their maximum likelihood estimates. These cannot be obtained explicitly, but require a computer. However, approximate moment estimates are $\hat{\mu} \doteq \bar{y}_\bullet$ and $\hat{\sigma}^2 \doteq s^2$.

Example (continued)
Let us apply this model to the data on the time to do homework assignments. These data are reproduced in Table 4.12, along with the fitted logistic distribution. As an approximation, we can use the same parameter estimates as for the normal distribution; the maximum likelihood estimates are $\hat{\mu} = 16.86$ and $\sigma = 8.19$.

Because of the similarity of the two distributions, we expect the same problems as for the normal distribution: the standardised residuals show the same pattern. However, the AIC is slightly better: 6.9 instead of 7.4 (7 for the multinomial), both with 5 d.f. The two are compared graphically in Figure 4.17.   □

Table 4.12. Times (hours) reported spent doing a homework assignment by the first class taking this course in 1976, with the fitted logistic distribution and the residuals.

Hours	Students	Multinomial	Logistic	Residual
5–9	3	0.115	0.088	0.477
10–14	8	0.308	0.197	1.266
15–19	6	0.231	0.284	−0.511
20–24	5	0.192	0.221	−0.311
25–29	1	0.038	0.105	−1.045
30–34	2	0.077	0.039	0.990
35–39	0	0.000	0.013	−0.581
40–44	1	0.038	0.004	2.686

Relationship to logistic regression   Because of its name, the reader may by now be wondering if there is a relationship between the logistic distribution and the logistic models of Chapter 2. The answer is yes. The logistic cumulative distribution function (Section 1.3.4) is

$$F(y_i; \mu, \sigma^2) = \frac{e^{-\frac{\pi}{\sqrt{3}}\left(\frac{y_i - \mu}{\sigma}\right)}}{1 + e^{-\frac{\pi}{\sqrt{3}}\left(\frac{y_i - \mu}{\sigma}\right)}} \tag{4.9}$$

Notice, among other things, that the square has disappeared in the denominator, as compared with the density function.

In order to interpret this, consider the following model. Suppose that we have some continuous variable x_i. As x_i increases or *cumulates* for an individual, a certain threshold or tolerance is passed. Then, the individual changes from some given state to another. However, the threshold is unobservable, so that only the state is recorded as the response variable. If these thresholds have a logistic distribution, then the probability π_i of change of state will be related to x_i by a logistic regression model:

$$\log\left(\frac{\pi_i}{1 - \pi_i}\right) = \beta_0 + \beta_1 x_i$$

$$\pi_i = \frac{e^{\beta_0 + \beta_1 x_i}}{1 + e^{\beta_0 + \beta_1 x_i}}$$

The latter is just the cumulative distribution function of Equation (4.9) with

$$\beta_0 = \frac{\pi \mu}{\sqrt{3}\sigma}$$

$$\beta_1 = -\frac{\pi}{\sqrt{3}\sigma}$$

An example was plotted in Figure 2.5.

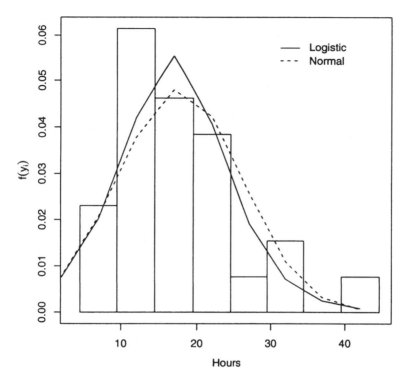

Fig. 4.17. The histogram for the number of hours taken to do a homework assignment, with the logistic and normal density functions.

4.4.3 LAPLACE DISTRIBUTION

A third symmetric distribution, related to the normal distribution, arises if we considered differences from the central value as absolute values $|y_i - \mu|$ instead of squares $(y_i - \mu)^2$. This is called the Laplace distribution

$$\pi_i = f(y; \mu, \sigma)\Delta_i$$
$$= \frac{e^{-\left|\frac{y_i - \mu}{\sigma}\right|}}{2\sigma}$$

An example curve is plotted in Figure 4.18. We can see that it is much narrower and more peaked than the normal distribution, but with thicker tails. This distribution is sometimes called the double exponential because the density has two exponential curves placed back to back.

Here, μ is not the mean but rather the *median* (Section 1.2.2). The theoretical variance is

$$\sigma_T^2 = 2\sigma^2$$

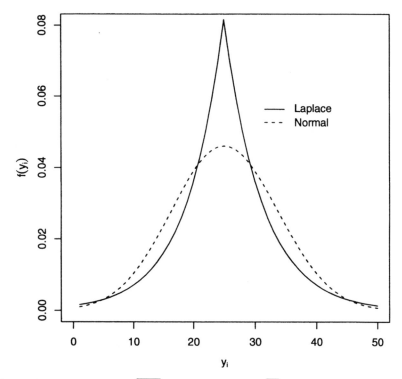

Fig. 4.18. Laplace ($\sigma = \sqrt{75/2}$) and normal ($\sigma = \sqrt{75}$) density functions, both with $\mu = 25$, so that both have the same variance.

This is another of the small group of distribution for which the maximum likelihood estimates can be obtained explicitly. As for the mean, with grouped data, the median can only be calculated approximately to give the maximum likelihood estimate for μ. As might be expected, the corresponding estimate for the dispersion parameter is

$$\hat{\sigma} = \frac{1}{n_\bullet} \sum_i n_i |y_i - \hat{\mu}|$$

This average sum of the absolute values of the deviations from the median is an alternative measure of dispersion instead of the average sum of the squared deviations from the mean. Because of the relationship between σ and the theoretical variance, we can expect the standard deviation to be $\sqrt{2}$ times larger than this dispersion parameter.

Table 4.13. Times (hours) reported spent doing a homework assignment by the first class taking this course in 1976, with the fitted Laplace distribution and the residuals.

Hours	Students	Multinomial	Laplace	Residual
5–9	3	0.115	0.080	0.638
10–14	8	0.308	0.180	1.530
15–19	6	0.231	0.406	−1.403
20–24	5	0.192	0.180	0.145
25–29	1	0.038	0.080	−0.749
30–34	2	0.077	0.036	1.120
35–39	0	0.000	0.016	−0.640
40–44	1	0.038	0.007	1.918

Example (continued)
The results when the Laplace distribution is applied to the data on the time to do homework assignments are shown in Table 4.13. The maximum likelihood estimates are $\hat{\mu} = 17$ and $\hat{\sigma} = 6.15$. The AIC, 6.8, is about the same as for the logistic distribution. The residuals in Table 4.13 show no particular pattern. The Laplace distribution is compared graphically with the normal distribution for these data in Figure 4.19. The third category, where the peak lies, is badly predicted. □

4.4.4 CAUCHY DISTRIBUTION

Both the logistic and the Laplace distributions were symmetric with thicker tails than the normal. Another distribution with still thicker tails is the Cauchy

$$\pi_i = f(y;\mu,\sigma)\Delta_i$$
$$= \frac{\Delta_i}{\pi\sigma\left[1+\left(\frac{y_i-\mu}{\sigma}\right)^2\right]}$$

This distribution has such thick tails that, in contrast to the logistic and Laplace distributions, both the theoretical mean and variance are undefined. Thus, μ describes the location of the distribution but is not the mean and σ describes the variability but is not the standard deviation. An example curve is plotted in Figure 4.20.

Because the theoretical mean and variance are not available, the approximate methods to obtain parameter estimates, used for some of the other distributions in this chapter, are not applicable here. A computer is required to obtain the maximum likelihood estimates.

Example (continued)
When the Cauchy distribution is applied to the data on the time to do homework assignments, the results are as shown in Table 4.14. The maximum likelihood

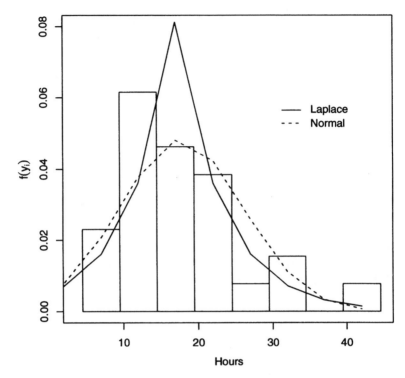

Fig. 4.19. The histogram for the number of hours taken to do a homework assignment, with the Laplace and normal density functions.

Table 4.14. Times (hours) reported spent doing a homework assignment by the first class taking this course in 1976, with the fitted Cauchy distribution and the residuals.

Hours	Students	Multinomial	Cauchy	Residual
5–9	3	0.115	0.077	0.703
10–14	8	0.308	0.218	0.975
15–19	6	0.231	0.319	−0.794
20–24	5	0.192	0.116	1.148
25–29	1	0.038	0.047	−0.210
30–34	2	0.077	0.025	1.695
35–39	0	0.000	0.015	−0.624
40–44	1	0.038	0.010	1.453

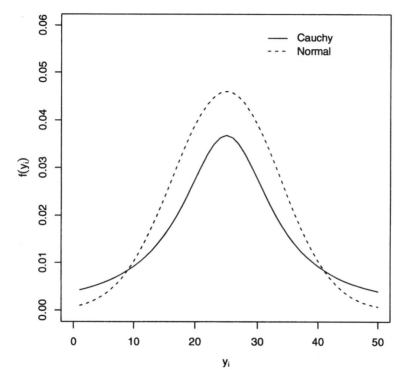

Fig. 4.20. Cauchy and normal density functions with $\mu = 25$ and $\sigma = \sqrt{75}$.

estimates are $\hat{\mu} = 15.54$ and $\hat{\sigma} = 4.52$. The AIC, 10.0, is worse than for the normal, logistic, and Laplace distributions. This may be because the Cauchy distribution predicts more negative times. However, the residuals in Table 4.14 show no particular pattern. The Cauchy distribution is compared graphically with the normal distribution for these data in Figure 4.21.   □

4.4.5 STUDENT t DISTRIBUTION

The Cauchy distribution may have tails that are too thick for a given response variable. It may predict too many 'outliers'. However, this distribution may be generalised by adding a parameter, say ν, to yield the Student t distribution

$$
\begin{aligned}
\pi_i &= f(y_i; \mu, \sigma, \nu)\Delta_i \\
&= \frac{\Gamma\left(\frac{\nu+1}{2}\right)\Delta_i}{\Gamma\left(\frac{\nu}{2}\right)\sigma\sqrt{\nu\pi}\left[1 + \frac{1}{\nu}\left(\frac{y_i - \mu}{\sigma}\right)^2\right]^{\frac{\nu+1}{2}}}
\end{aligned}
$$

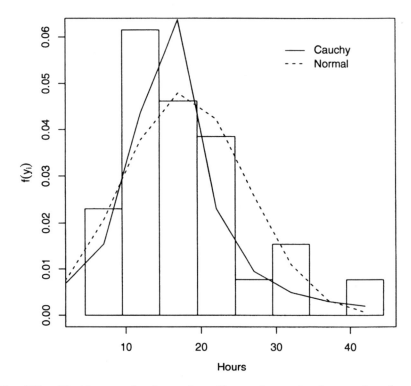

Fig. 4.21. The histogram for the number of hours taken to do a homework assignment, with the Cauchy and normal density functions.

When $\nu = 1$, the distribution is Cauchy and, when it is infinite, the distribution is normal. For values in between, the tails vary in thickness between these two distributions. Figure 4.22 provides a comparison of the Student t with $\nu = 5$ and its two special cases, the Cauchy and normal distributions.

The theoretical mean is only finite if $\nu > 1$. The theoretical variance is

$$\sigma_T^2 = \frac{\nu\sigma^2}{\nu - 2}, \qquad \nu > 2$$

not existing otherwise. As with the Cauchy distribution, the parameter estimates can be obtained by maximum likelihood using a computer.

Example (continued)
We can now see if the Student t distribution applied to the data on the time to do homework assignments provides a better fit than the previous distributions. The results are shown in Table 4.15. The maximum likelihood estimates are $\hat{\mu} = 16.63$, $\hat{\sigma} = 6.59$, and $\hat{\nu} = 5.03$. The AIC is 7.9, about the same as for the logistic

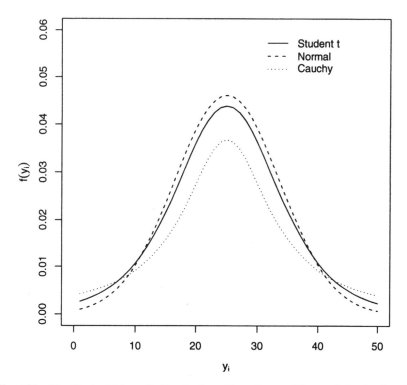

Fig. 4.22. The Student t ($\nu = 5$), Cauchy ($\nu = 1$), and normal ($\nu = \infty$) density functions with $\mu = 25$ and $\sigma = \sqrt{75}$.

Table 4.15. Times (hours) reported spent doing a homework assignment by the first class taking this course in 1976, with the fitted Student t distribution and the residuals.

Hours	Students	Multinomial	Student t	Residual
5–9	3	0.115	0.099	0.265
10–14	8	0.308	0.217	0.991
15–19	6	0.231	0.288	−0.540
20–24	5	0.192	0.198	−0.069
25–29	1	0.038	0.086	−0.830
30–34	2	0.077	0.032	1.299
35–39	0	0.000	0.012	−0.550
40–44	1	0.038	0.005	2.547

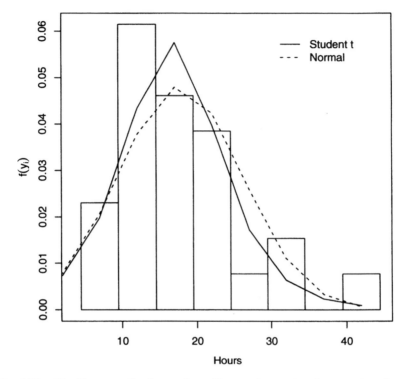

Fig. 4.23. The histogram for the number of hours taken to do a homework assignment, with the Student t ($\hat{\nu} = 5.03$) and normal density functions.

and Laplace distributions, but better than the Cauchy. (Remember that the penalty in the AIC involves an extra parameter so that there is one less degree of freedom.) Again, the residuals in Table 4.15 are similar to those for the normal and logistic distributions, not as extreme as some of those for the Laplace. The Student t distribution is compared graphically with the normal distribution for these data in Figure 4.23.

We can conclude that these data exhibit no indication of outliers, at least ac-cording to the Student t (and Cauchy) model.   □

As we shall see in Chapter 5, in certain contexts, this distribution is used instead of the Chi-squared distribution (Section 3.4.3) for making frequentist in-ferences. In that case, the parameter ν, with integer values, indicates the number of degrees of freedom.

4.5 Distributions for durations

Probability distributions that describe duration times until some event occurs are another important class of models. Depending on the field in which they are used, these are variously called lifetime, survival, failure, loss, or reliability distributions. We have already seen one example in Section 4.3.2: the geometric distribution, which uses equal discrete increments of time. Those that I shall look at here are for continuous time. However, these distributions are much more widely applicable; they can be used for any response variable that is restricted to have only positive values.

Distributions suitable for modelling durations and other positive-valued responses are characterised by their form:

- durations can only be positive response variables;
- shorter durations are usually most probable, but with small probabilities tailing off for longer times.

One must wait a *random* time in order to observe each event. Unfortunately, such a wait, if lengthy, may not always be possible, for reasons of cost or time limitations of the study. Thus, the observed response of some individuals may only be that the event has not yet occurred before observation had to stop. Such an observation is said to be *censored*. Such data still contain useful information: that the duration has at least the observed length. I shall look at censored data in more detail in Sections 4.5.2 and 6.4.

Duration distributions have two major areas of application.

(1) A sample of individuals may be observed, with one event (possibly) occurring to each, and the duration recorded.

(2) One or more individuals may be observed over an extended period and the times of a series of repeatable events recorded.

In the first case, the event is often terminal, called *absorbing*.

Examples
Death, recovery from a serious illness, and marriage are all possible events signifying end of a duration for each individual. The first is certainly absorbing!   □

More complex applications of the second case occur when a series of different events is studied.

Examples
In medical applications, a series of repeatable events might involve epileptic fits, migraines, infections, or asthma attacks. A series of different events could be becoming seriously ill, entering hospital, and recovering or dying.

In social applications, repeated events might be changing jobs. Different events could be graduating, finding a job, and obtaining a promotion.   □

Naturally, censoring is more frequent when one event is to be recorded per individual than when a series of events is studied. (Why?)

4.5.1 INTENSITY AND SURVIVOR FUNCTIONS

In previous sections, we have seen that the probability density function is central to modelling response variables. Although this remains the case with durations, two other related functions also play key roles.

Intensity function   An important function associated with duration distributions is the intensity $\lambda(y_i)$ of events. It describes the rate at which events are occurring at different points in time y_i. We have already encountered one such function in connection with the Poisson process in Section 4.3.1, where it was constant over time. However, in most realistic situations, the intensity will change over time.

Examples
If there is ageing, the risk will increase over time. If there is an initial wearing-in period, it will be high at the beginning, later decreasing.   □

Each duration distribution which we shall study has a different intensity function, depending on time, although some cannot be written down as simple functions.

Survivor function   A second frequently used function is the (cumulative) survival probability, one minus the cumulative distribution function of Equation (1.11). This is called the survivor function; it gives the probability of a duration lasting at least the given time y:

$$S(y_i) = \Pr(Y > y_i)$$
$$= 1 - F(y_i)$$

where $F(y_i)$ is the cumulative distribution function (Section 1.3.4). The survivor function, instead of the density, gives the appropriate probability for a censored observation. (Why?)

Function relationships   The following relationship holds among the density, intensity, and survivor functions:

$$\lambda(y_i)\Delta_i = \frac{f(y_i)\Delta_i}{S(y_i)}$$

The intensity gives the conditional probability of an event occurring at time y_i, and the duration ending then, given that it has not yet happened. Notice that this is a special kind of conditional probability because the condition is not fixed. Usually (Section 1.3.2), a conditional probability of one variable is for a *fixed* value of some other variable. Here, it depends on the *same* variable so that the condition changes with y_i.

For many distributions, $F(y_i)$ involves a complex integral so that the survivor and intensity functions are not available in explicit form.

Examples
The normal cumulative distribution function (Section 4.4.1) involves an integral so that the survivor and intensity functions are not explicitly available.

In contrast, the logistic cumulative distribution function (Section 4.4.2) can be explicitly written down. Thus, from Equation (4.9), the survivor function is

$$S(y_i;\mu,\sigma^2) = 1 - F(y_i;\mu,\sigma^2)$$
$$= 1 - \frac{e^{-\frac{\pi}{\sqrt{3}}\left(\frac{y_i-\mu}{\sigma}\right)}}{1+e^{-\frac{\pi}{\sqrt{3}}\left(\frac{y_i-\mu}{\sigma}\right)}}$$
$$= \frac{1}{1+e^{-\frac{\pi}{\sqrt{3}}\left(\frac{y_i-\mu}{\sigma}\right)}}$$

and, by dividing the density in Equation (4.8) by this, the intensity function is

$$\lambda(y_i;\mu,\sigma) = \frac{\pi e^{-\frac{\pi}{\sqrt{3}}\left(\frac{y_i-\mu}{\sigma}\right)}}{\sqrt{3}\sigma\left[1+e^{-\frac{\pi}{\sqrt{3}}\left(\frac{y_i-\mu}{\sigma}\right)}\right]}$$

However, as they stand, these distributions are not really appropriate for modelling durations because they are symmetric and they give positive probability to negative duration times. (See, however, Section 4.6.1.) □

4.5.2 EXPONENTIAL DISTRIBUTION
The simplest duration distribution is the exponential with density function

$$\pi_i = f(y_i;\mu)\Delta_i$$
$$= \lambda e^{-\lambda y_i}\Delta_i, \qquad \lambda > 0$$
$$= \frac{1}{\mu}e^{-\frac{y_i}{\mu}}\Delta_i, \qquad \mu > 0$$

(The Laplace distribution of Section 4.4.3 could be constructed as two of these functions, back to back, hence its alternative name, double exponential.) The corresponding survivor function is

$$S(y_i;\mu) = e^{-\lambda y_i}$$
$$= e^{-\frac{y_i}{\mu}}$$

so that, by division, the intensity $\lambda = 1/\mu$ is constant, not depending on y_i.

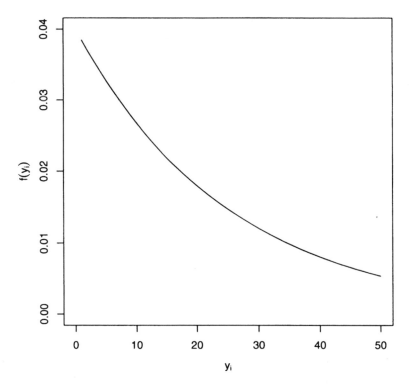

Fig. 4.24. The exponential density function with $\mu = 25$ or $\lambda = 0.04$.

Here, μ is the (theoretical) mean duration time between events, and its reciprocal, λ, is the *constant* intensity of events. The theoretical variance is the square
of the mean, $\sigma_T^2 = \mu^2$, so that it increases rapidly with the mean duration. Because the intensity is constant, the probability of an event is independent of the
duration already passed since the start of the period. This is, again, the memory-
less Markov property. The exponential distribution is the continuous analogue of
the geometric distribution. An example is plotted in Figure 4.24, showing that it
has a monotonely decreasing density.

If no observations are censored, the mean has the usual maximum likelihood
estimate, $\hat{\mu} = \bar{y}_\bullet$, and the estimate of the intensity is its reciprocal. If some observations are censored, the likelihood function is the product of densities for
the uncensored observations and survivor functions for those censored (Section
4.5.1):

$$L(\lambda) = \prod_i \left[\lambda e^{-\lambda y_i} \Delta_i\right]^{n_i} \prod_j \left[e^{-\lambda y_j} \Delta_j\right]^{n_j}$$

$$\propto \lambda^m e^{-\lambda y_\bullet}$$

Table 4.16. Days between coal-mining disasters in Great Britain, 1851–1962, with the fitted exponential distribution. (Lindsey, 1992, pp. 66–67, from Jarrett, 1979)

Days	Disasters	Multinomial	Exponential	Residual
0–20	28	0.147	0.089	2.668
20–40	20	0.105	0.081	1.150
40–60	17	0.089	0.074	0.775
60–80	11	0.058	0.068	−0.511
80–100	14	0.074	0.061	0.678
100–120	6	0.032	0.056	−1.422
120–140	13	0.068	0.051	1.065
140–200	17	0.089	0.127	−1.444
200–260	16	0.084	0.096	−0.512
260–320	11	0.058	0.072	−0.735
320–380	13	0.068	0.055	0.821
380–440	3	0.016	0.041	−1.723
440–500	4	0.021	0.031	−0.783
> 500	17	0.089	0.098	3.073

where i indexes uncensored observations, j indexes censored times, and $m = \sum_i n_i$ is the number of uncensored observations. Then, the maximum likelihood estimate of the mean duration is

$$\hat{\mu} = \frac{y_\bullet}{m}$$

which is larger than $\bar{y}_\bullet = y_\bullet/n_\bullet$ because the censored durations are actually longer than the observed times y_j. The denominator m is the number of *events* recorded; it does not include the number of incomplete durations, although the numerator does. Thus, the constant intensity will be estimated to be $\hat{\lambda} = m/y_\bullet$.

Example

Table 4.16 gives the durations in days between coal-mining disasters in Great Britain. They concern explosions in mines involving more than 10 men killed, as originally recorded in the *Colliery Year Book and Coal Trades Directory* produced by the National Coal Board in London, UK. This is an example of a series of repeated events.

From the original uncensored data, measured to the nearest day, the mean is estimated to be $\hat{\mu} = 213.4$ days. On the other hand, if we estimate the mean from the grouped observations in Table 4.16, using the formula for censored data with the last category given the censoring value 500, we obtain 185.1. (Note that $\bar{y}_\bullet = 168.6$ is even worse.) This is because there are some extremely long durations near the end of the series, with a maximum of 2366 days.

Because of the censoring in the last category of the table, the estimated probability in that category can be calculated as one minus the sum of the others.

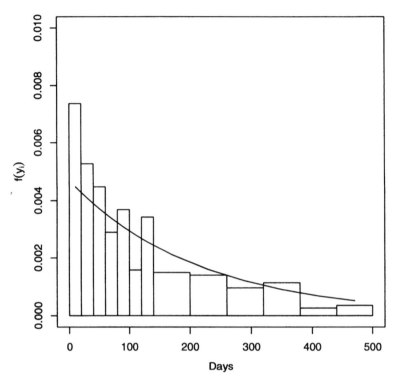

Fig. 4.25. The histogram and exponential density function for the times between mine disasters.

The estimated density function is plotted with the histogram in Figure 4.25. The AIC is 11.2 as compared with 13 for the multinomial, with 12 d.f, indicating a reasonable fit. The residuals in Table 4.16 do not show any clear pattern.

Closer inspection of the original data, however, indicates that the intensity of accidents may have changed about 1890. This was probably due to changes in the mining safety legislation around that time, or could be related to man-hours worked or tons of coal produced. The data are separated into these two periods in Table 4.17. The estimated means, again from the original data, are now, respectively, 117.0 and 398.8 days, showing the great change in average duration between disasters. The estimated density function for the first period (1851–1891) is plotted with the corresponding histogram in Figure 4.26. The AICs are respectively 5.4 and 12.8, again as compared with 13 for the multinomial. However, the apparently poorer fit in the second period may partially be due to the heavy grouping of the last category.

Thus, there is little evidence against an exponential distribution, both for the complete data, and separately for each period. However, by looking at them sep-

Table 4.17. Days from Table 4.16, separated into two periods, 1851–1891 and 1891–1962, with the fitted exponential distributions.

Time	1851–1891			1891–1962		
	Disasters	Mult.	Exp.	Disasters	Mult.	Exp.
0–20	22	0.177	0.152	6	0.092	0.049
20–40	17	0.137	0.129	3	0.046	0.047
40–60	13	0.105	0.109	4	0.062	0.044
60–80	8	0.065	0.093	3	0.046	0.042
80–100	14	0.113	0.079	0	0.000	0.040
100–120	6	0.048	0.067	0	0.000	0.038
120–140	9	0.073	0.056	4	0.062	0.036
140–200	12	0.097	0.122	5	0.077	0.098
200–260	11	0.089	0.074	5	0.077	0.085
260–320	4	0.032	0.045	7	0.108	0.073
320–380	4	0.032	0.027	9	0.138	0.063
380–440	2	0.016	0.017	1	0.015	0.054
440–500	0	0.000	0.010	4	0.062	0.046
> 500	2	0.016	0.020	15	0.231	0.286

arately, we have discovered an important change in intensity. □

Relationship to the Poisson distribution In the Poisson process of Section 4.3.1, the numbers of events in fixed intervals of time have a Poisson distribution and the intensity of such events is constant over time. Then, the exponential distribution, with its constant intensity, describes the lengths of time between the events in such a process.

The parameter μ is not the same for the two distributions: for the Poisson distribution, it is the mean number of events in a fixed time interval Δt, whereas, for the exponential distribution, it is the mean length of time between events. In fact, for a unit time interval, the one is the reciprocal of the other. Thus, the intensity can be estimated either from the durations between events, as λ in the exponential distribution, or from the frequency of events, as the μ in the Poisson distribution.

As we see, with an exponential distribution we are assuming a constant intensity function over all time. As well, the density function has a rigid, monotonely decreasing form. In the next two sections, I shall consider two more flexible models built upon this distribution.

4.5.3 WEIBULL DISTRIBUTION

The first distribution based on the exponential is the Weibull. It can be interpreted as if several unobservable processes are running in parallel, each following an exponential distribution. Then, the first to stop ends the observable duration. This is a weakest link mechanism.

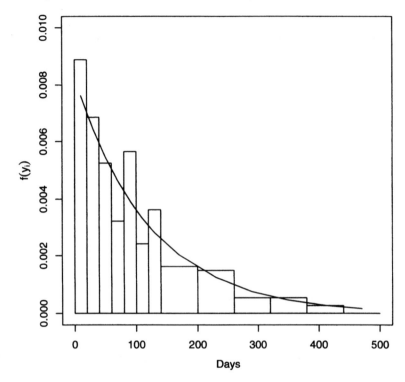

Fig. 4.26. The histogram and exponential density function for the times between mine disasters for the period 1851–1891.

Example
Most machines consist of a large number of parts. The failure of some part, the weakest link, may cause the machine to break down. Then, the observed duration is the total operating time of the machine. Interestingly, Weibull was an engineer. ☐

The Weibull density function is

$$\pi_i = f(y_i; \alpha, \mu)\Delta_i$$
$$= \frac{\alpha y_i^{\alpha-1} e^{-(y_i/\mu)^\alpha}}{\mu^\alpha}\Delta_i, \qquad \mu, \alpha > 0$$

Plots of two Weibull distributions are given in the left graph of Figure 4.27. We see that the density can either be continuously decreasing or have a mode. The corresponding survivor function is

$$S(y_i; \alpha, \mu) = e^{-(y_i/\mu)^\alpha}$$

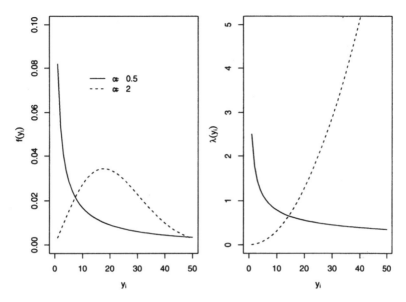

Fig. 4.27. Weibull density functions with $\mu = 25$ and $\alpha = 0.5$ and 2 and the corresponding intensity functions.

and the intensity function

$$\lambda(y_i; \alpha, \mu) = \frac{\alpha y_i^{\alpha-1}}{\mu^\alpha}$$

The latter increases without limit for $\alpha > 1$ and decreases towards zero for $\alpha < 1$. This can be seen in the right graph of Figure 4.27. When $\alpha = 1$, these functions reduce to the corresponding ones for an exponential distribution.

The theoretical mean and variance are, respectively,

$$\mu_T = \mu \Gamma\left(1 + \frac{1}{\alpha}\right)$$

$$\sigma_T^2 = \mu^2 \left[\Gamma\left(1 + \frac{2}{\alpha}\right) - \Gamma\left(1 + \frac{1}{\alpha}\right)^2\right]$$

Thus, both maximum likelihood and moment estimates are difficult to calculate. An approximate estimate of α can be obtained from

$$\frac{\bar{y}_\bullet^2}{s^2 + \bar{y}_\bullet^2} \doteq \frac{\Gamma\left(1 + \frac{1}{\hat{\alpha}}\right)^2}{\Gamma\left(1 + \frac{2}{\hat{\alpha}}\right)}$$

Values are listed in Table A.6 of the Appendix, from which the approximate parameter estimate can be obtained after calculating the left hand side of this equation. (However, if this ratio is close to zero or one, that is, if either $\bar{y}_\bullet$ or s^2 is

close to zero, then the estimate is not very accurately obtainable, and has little
meaning.) Then,

$$\hat{\mu} \doteq \frac{\bar{y}_{\bullet}}{\Gamma\left(1 + \frac{1}{\hat{\alpha}}\right)}$$

On the other hand, maximum likelihood estimation requires a computer.

Example

Let us look at the lengths of marriage for all couples divorcing in Liège, Belgium, in 1984. This is an example of many individuals each having one event.
However, they are unusual for duration data in that they are retrospective, with
all marriages ending at the same time. Here, the Weibull mechanism could be
the 'last straw' that breaks the marriage! Table 4.18 gives the data and the fitted
Weibull distribution.

The empirical mean and variance are, respectively, 13.85 years and 75.90.
From these estimates, we obtain a value of 0.717 to look up in Table A.6. This
gives $\hat{\alpha} \doteq 1.63$ by interpolation. Then, we have $\hat{\mu} \doteq 15.48$. In comparison, the
maximum likelihood estimates are $\hat{\mu} = 15.60$ and $\hat{\alpha} = 1.69$. The AIC is 86.1 as
compared with 51 for the multinomial so that the fit is not good, although there are
49 d.f. This can also be seen in Figure 4.28. The Weibull curve is flatter and more
dispersed than the histogram. However, $n_{\bullet} = 1699$, many more observations than
in most previous examples.   □

4.5.4 GAMMA DISTRIBUTION

Now suppose that the observable duration is made up of the sum of a series of
several unobservable or latent periods, each having a different but constant intensity of an exponential distribution. Then, the total duration will have a gamma
distribution,

$$\begin{aligned}
\pi_i &= f(y_i; \alpha, \mu)\Delta_i \\
&= \frac{\alpha^\alpha y_i^{\alpha-1} e^{-\alpha y_i/\mu}}{\mu^\alpha \Gamma(\alpha)} \Delta_i \qquad \mu, \alpha > 0
\end{aligned}$$

with $\Gamma(\cdot)$ again the gamma function. Two examples are plotted in Figure 4.29. As
with the Weibull distribution, we see that the density can either be continuously
decreasing or have a mode.

When α is an integer, this parameter is the number of latent periods making
up the duration. The distribution is then sometimes called the Erlang distribution.
The gamma distribution is the continuous analogue of the negative binomial.

Because of the integral involved, the gamma intensity function is rather complicated. However, it increases to a maximum of α/μ for $\alpha > 1$ and decreases to
a minimum at that same value for $\alpha < 1$. This contrasts with the Weibull intensity which increases without limit or decreases to zero. This difference can be of
theoretical use in choosing between the two distributions.

Table 4.18. Length of marriage (years) before divorce in Liège, 1984, with the fitted Weibull distribution. (Lindsey, 1992, pp. 14–15)

Years	Divorces	Mult.	Weibull	Years	Divorces	Mult.	Weibull
1	3	0.002	0.019	27	14	0.008	0.012
2	18	0.011	0.028	28	17	0.010	0.011
3	59	0.035	0.035	29	12	0.007	0.010
4	87	0.051	0.040	30	17	0.010	0.008
5	82	0.048	0.044	31	10	0.006	0.007
6	90	0.053	0.047	32	11	0.006	0.006
7	91	0.054	0.049	33	13	0.008	0.005
8	109	0.064	0.050	34	7	0.004	0.005
9	94	0.055	0.050	35	9	0.005	0.004
10	83	0.049	0.049	36	9	0.005	0.003
11	101	0.059	0.048	37	9	0.005	0.003
12	91	0.054	0.046	38	10	0.006	0.002
13	94	0.055	0.044	39	5	0.003	0.002
14	63	0.037	0.042	40	3	0.002	0.002
15	68	0.040	0.040	41	3	0.002	0.001
16	56	0.033	0.037	42	4	0.002	0.001
17	62	0.036	0.035	43	6	0.004	0.001
18	40	0.024	0.032	44	0	0.000	0.001
19	43	0.025	0.030	45	0	0.000	0.001
20	41	0.024	0.027	46	1	0.001	0.001
21	28	0.016	0.025	47	0	0.000	0.000
22	24	0.014	0.022	48	2	0.001	0.000
23	39	0.023	0.020	49	0	0.000	0.000
24	34	0.020	0.018	50	0	0.000	0.000
25	14	0.008	0.016	51	0	0.000	0.000
26	22	0.013	0.014	52	1	0.001	0.000

The theoretical mean is μ and the theoretical variance

$$\sigma_T^2 = \frac{\mu^2}{\alpha}$$

Thus, the parameter α is the ratio of the mean squared to the variance. The square root of the reciprocal of this, the ratio of the standard deviation to the mean, is called the *coefficient of variation*. Like the coefficient of dispersion for count data, this provides a standard for measuring variability. When $\alpha = 1$, this distribution reduces to the exponential, which thus has a coefficient of variation of one.

Because the maximum likelihood estimate of α is difficult to calculate without a computer, it can be approximated by the moment estimate, $\hat{\alpha} \doteq \bar{y}_\bullet^2 / s^2$, using $\hat{\mu} = \bar{y}_\bullet$.

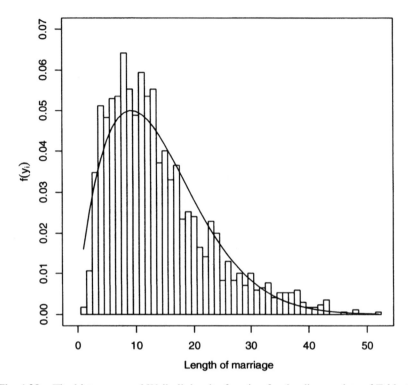

Fig. 4.28. The histogram and Weibull density function for the divorce data of Table 4.18.

The gamma distribution has a reproductive property. Suppose that Y_1 and Y_2 follow gamma distributions with different values of α, say α_1 and α_2, but the same value of μ. Then, the sum $Y_1 + Y_2$ will also follow a gamma distribution, but with parameters $\alpha_1 + \alpha_2$ and 2μ.

Example (continued)
Marriage is sometimes theorised as having three periods of differing relationships through which a couple passes before a divorce occurs, so that a gamma distribution with $\alpha \doteq 3$ might be suitable. Let us fit this distribution to the divorce data of Table 4.18. These are reproduced in Table 4.19, with the fitted distribution.

The parameter estimates are $\hat{\mu} = 13.85$ and $\hat{\alpha} \doteq 2.53$, with maximum likelihood estimate $\hat{\alpha} = 2.65$; the latter is not too far from the theoretical value cited above. The AICs are 51.8 and 51 for the gamma and multinomial, respectively, indicating a reasonable fit, considering the number of observations and the 49 d.f. The fit certainly is an improvement on that for the Weibull distribution. The histogram and fitted gamma distribution are plotted in Figure 4.30 and appear reasonably close. For comparison, the Weibull distribution is also shown.

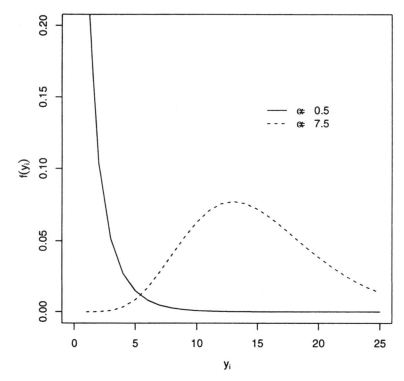

Fig. 4.29. Gamma density functions with $\mu = 1$, $\alpha = 0.5$ and $\mu = 15$, $\alpha = 7.5$ or χ^2 distributions with 1 and 15 d.f.

In contrast to the homework data (which are also duration data) in Section 4.4, the sample size for these data is large enough to provide power to distinguish among rather more complex models than two parameter ones like the gamma and Weibull distributions. However, if no such alternative is available, one would be prepared to accept provisionally the three-stage mechanism of the gamma model instead of the 'last straw' mechanism of the Weibull. □

When $\mu/\alpha = 2$, the gamma distribution is a χ^2 distribution with 2α degrees of freedom, as was used in Section 3.4.3. Thus, Figure 4.29 shows two such χ^2 distributions with 1 and 15 d.f.

4.5.5 INVERSE GAUSS DISTRIBUTION

Let us now develop still another model for duration times. Suppose that some unobservable (latent) process u_t is evolving over time t. The duration until such a process reaches, for the first time, some fixed value or boundary, say b, greater than the initial value, say u_0, is called the *hitting* or *waiting time*.

Table 4.19. Length of marriage (years) before divorce in Liège, 1984, with the fitted gamma distribution. (Lindsey, 1992, pp. 14–15)

Years	Divorces	Mult.	Gamma	Years	Divorces	Mult.	Gamma
1	3	0.002	0.008	27	14	0.008	0.011
2	18	0.011	0.020	28	17	0.010	0.010
3	59	0.035	0.031	29	12	0.007	0.009
4	87	0.051	0.040	30	17	0.010	0.008
5	82	0.048	0.047	31	10	0.006	0.007
6	90	0.053	0.052	32	11	0.006	0.006
7	91	0.054	0.055	33	13	0.008	0.005
8	109	0.064	0.056	34	7	0.004	0.004
9	94	0.055	0.056	35	9	0.005	0.004
10	83	0.049	0.054	36	9	0.005	0.003
11	101	0.059	0.052	37	9	0.005	0.003
12	91	0.054	0.050	38	10	0.006	0.003
13	94	0.055	0.047	39	5	0.003	0.002
14	63	0.037	0.044	40	3	0.002	0.002
15	68	0.040	0.041	41	3	0.002	0.002
16	56	0.033	0.037	42	4	0.002	0.001
17	62	0.036	0.034	43	6	0.004	0.001
18	40	0.024	0.031	44	0	0.000	0.001
19	43	0.025	0.028	45	0	0.000	0.001
20	41	0.024	0.025	46	1	0.001	0.001
21	28	0.016	0.023	47	0	0.000	0.001
22	24	0.014	0.020	48	2	0.001	0.001
23	39	0.023	0.018	49	0	0.000	0.001
24	34	0.020	0.016	50	0	0.000	0.000
25	14	0.008	0.014	51	0	0.000	0.000
26	22	0.013	0.013	52	1	0.001	0.000

Next, let us assume that successive changes in this process have a normal distribution,

$$u_t - u_{t-1} \sim N(1/\mu, \sigma^2), \qquad \mu > 0$$

This is called *Brownian motion* or a *Wiener process* with positive *drift*, because the mean is greater than zero. A typical graph of such a process is given in Figure 4.31. In this graph, $u_0 = 0$ and the hitting time occurs when u_i crosses the boundary $b = 20$ for the first time between $t = 20$ and 21.

Example

Workers might accumulate small random complaints about their job over time, the latent process. When these reach a certain critical level, individual workers

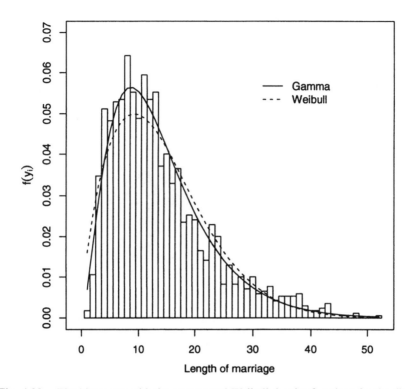

Fig. 4.30. The histogram with the gamma and Weibull density functions for the divorce data of Table 4.18.

change jobs or, collectively, a strike breaks out. The waiting or hitting time is the time between job changes or strikes.   □

Suppose now that this waiting time, a duration of some phenomenon, is the observed response variable, $y_i = t$. Under the conditions of these hypotheses, it will have an inverse Gauss distribution,

$$\pi_i = f(y_i; \mu, \sigma^2)\Delta_i$$

$$= \frac{b}{\sqrt{2\pi y_i^3 \sigma^2}} e^{-\frac{(y_i - b\mu)^2}{2y_i \sigma^2 \mu^2}} \Delta_i$$

Because u_t is unobservable, b is typically unknown and must be absorbed into the mean, or, equivalently, set to one. An example, with the same parameter values as for Figure 4.31, is plotted in Figure 4.32.

In this parametrisation, σ^2 does not represent the variance of the waiting time. Instead, the theoretical variance is

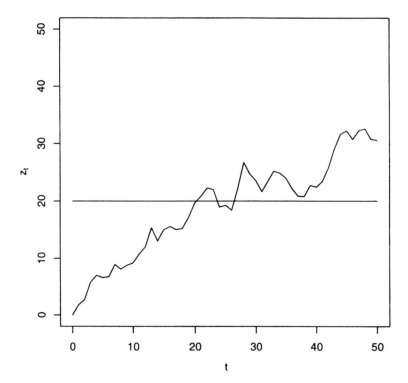

Fig. 4.31. Brownian motion starting at $u_0 = 0$, with mean, $1/\mu = 0.5$ and variance, $\sigma^2 = 4$.

$$\sigma_T^2 = \mu^3 \sigma^2$$

The maximum likelihood estimates are $\hat{\mu} = \bar{y}_\bullet$ and

$$\hat{\sigma}^2 = \frac{1}{n_\bullet} \sum_i n_i y_i \left(\frac{1}{y_i} - \frac{1}{\hat{\mu}} \right)^2$$

$$= \frac{1}{n_\bullet} \sum_i \frac{n_i}{y_i} - \frac{1}{\hat{\mu}}$$

Example (continued)
Because the times to do homework assignments are durations, the inverse Gauss distribution may be more appropriate than the symmetric distributions tried above. The latent process might be the students' hesitations until they find solutions to all the exercises! With this distribution, in contrast to the symmetric ones, negative times are impossible, having zero probability. Let us try it.

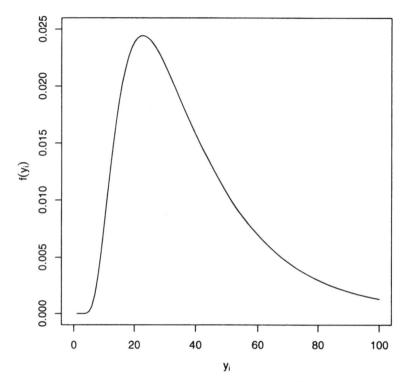

Fig. 4.32. Inverse Gauss density with, $b = 20$, $\mu = 2$ and $\sigma^2 = 4$ corresponding to the process in Figure 4.31.

The data are reproduced again in Table 4.20, with the fitted inverse Gauss distribution. The estimated density function for the inverse Gauss distribution is plotted with the histogram in Figure 4.33. The parameters are estimated as $\hat{\mu} = 17.48$ and $\hat{\sigma}^2 = 0.013$. Again, the standardised residuals are small. The AIC is 4.1 as compared with 7 for the multinomial, considerably better than any of the previously fitted models. Thus, we have evidence against the normal and other symmetric distributions, here in favour of the inverse Gauss. □

4.6 Transforming the response

Before the computer era, transformations to normality were widely applied in statistics. As we have seen, however, a wide variety of more appropriate non-normal distributions is available. In a modelling context, transformation only makes sense for continuous response variables. Such a transformation of a response variable always results in a change of *shape* of the distribution of the originally observed variable. Thus, this procedure can be used to create new dis-

Table 4.20. Times (hours) reported spent doing a homework assignment by the first class taking this course in 1976, with the fitted inverse Gauss distribution and the residuals.

Hours	Students	Multinomial	Inverse Gauss	Residual
5–9	3	0.115	0.146	−0.405
10–14	8	0.308	0.303	0.047
15–19	6	0.231	0.240	−0.096
20–24	5	0.192	0.147	0.610
25–29	1	0.038	0.081	−0.764
30–34	2	0.077	0.043	0.829
35–39	0	0.000	0.023	−0.765
40–44	1	0.038	0.012	1.268

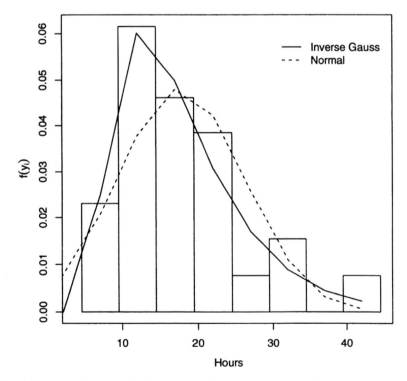

Fig. 4.33. The histogram for the number of hours taken to do a homework assignment, with inverse Gauss and normal density functions.

tributions with more appropriate shapes for the data. Almost all transformations require that the responses only take strictly positive values.

After applying a transformation, one must take great care in interpreting the parameters of the distribution. They will no longer refer to the original values of the response and will often be difficult to relate to the phenomenon under study.

4.6.1 LOG TRANSFORMATION

The most common transformation is obtained by taking logarithms of the values of the response variable. The original density must then be multiplied by $1/y_i$. (This is called the Jacobian and arises from the derivative of the logarithm transformation.)

Log normal distribution The most common case of log transformation is its application to the normal distribution, yielding the log normal distribution,

$$
\begin{aligned}
\pi_i &= f(y_i; \mu, \sigma^2)\Delta_i \\
&= \frac{1}{y_i\sqrt{2\pi\sigma^2}} e^{-\frac{1}{2\sigma^2}[\log(y_i)-\mu]^2} \Delta_i
\end{aligned}
$$

Notice that we cannot simply replace y_i by $\log(y_i)$ in Equation (4.5): the function is also divided by y_i. As well, in contrast to the normal distribution, here the responses must all be strictly positive.

The simplicity of this distribution results from the fact that normal distribution techniques can be applied to the logarithm of the response variable. Thus, μ and σ^2 are the mean and variance of the log response, and their maximum likelihood estimates are obtained in the same way as for the normal distribution, but calculated for the log response.

However, the interpretation of the results can be very difficult because the parameters have changed meaning. The theoretical mean and variance of the original response variable are *not* given by simple back transformation. They are more complex, respectively,

$$
\begin{aligned}
\mu_T &= e^{\mu+\sigma^2/2} \\
\sigma_T^2 &= e^{2\mu+\sigma^2}(e^{\sigma^2}-1) \\
&= \mu_T^2(e^{\sigma^2}-1)
\end{aligned}
$$

Indeed, e^μ is the geometric, not the arithmetic, mean of the original, untransformed responses. Notice that, in contrast to the normal distribution (see Figure 4.14), the theoretical variance of the log normal distribution depends on the theoretical mean. The coefficient of variation is $\sqrt{e^{\sigma^2}-1}$ so that there is overdispersion (with respect to an exponential distribution) if $\sigma^2 > \log(2)$.

This distribution is frequently used for positive-valued response variables, especially in economics. An example is plotted in Figure 4.34. Because the arith-

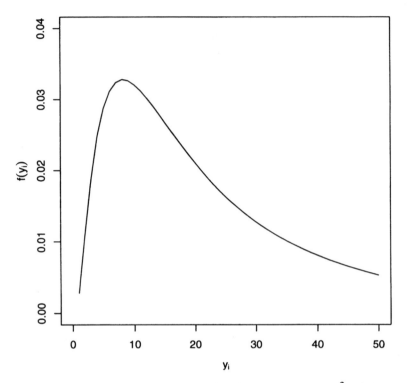

Fig. 4.34. The log normal density function with $\mu = 3$ and $\sigma^2 = 0.9$.

metic mean of the log normal distribution depends on the variance, the form of
the distribution changes when this mean changes, like the duration distributions of
Section 4.5. The distribution describes a continuous response variable that is the
result of a large number of small accumulating, unknown, multiplicative factors.

Example (continued)
We can apply this distribution to the data on the time to do homework assignments, reproduced in Table 4.21, with the fitted log normal distribution. Here,
the parameters are estimated as $\hat{\mu} = 2.77$ and $\hat{\sigma}^2 = 0.208$, so that the theoretical mean homework time is estimated as $e^{2.774+0.208/2} = 17.79$. The estimated
density function for the log normal distribution is plotted with the histogram in
Figure 4.35. As with the inverse Gauss distribution, negative times are impossible, having zero probability. The standardised residuals are smaller and vary
rather randomly, as compared with those for the normal distribution above. The
AIC is 4.2 compared with 7 for the multinomial. The results are very close to
those for the inverse Gauss distribution.

Thus, we have further evidence against symmetric distributions such as the

Table 4.21. Times (hours) reported spent doing a homework assignment by the first class taking this course in 1976, with the fitted log normal distribution and the residuals.

Hours	Students	Multinomial	Log normal	Residual
5–9	3	0.115	0.134	−0.262
10–14	8	0.308	0.300	0.070
15–19	6	0.231	0.248	−0.179
20–24	5	0.192	0.151	0.547
25–29	1	0.038	0.082	−0.769
30–34	2	0.077	0.042	0.859
35–39	0	0.000	0.022	−0.752
40–44	1	0.038	0.011	1.310

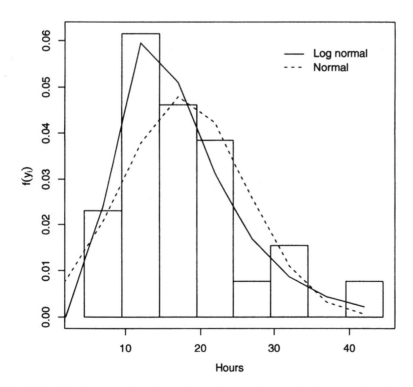

Fig. 4.35. The histogram for the number of hours taken to do a homework assignment, with log normal and normal density functions.

normal, in favour of some skewed distribution, such as the log normal or inverse Gauss. However, the number of observations is too small to have enough power really to distinguish empirically between the log normal and inverse Gauss distributions. Several other distributions in Section 4.5 would probably also be plausible for these data. Arguments for one or the other might be based on the relative plausibility of the data generating mechanisms that they describe.   □

Other symmetric distributions   If only positive responses are possible and a distribution with heavy tails is required, a log transformation can be applied to one of the other symmetric distributions of Section 4.4. Thus, the log logistic, log Laplace, log Cauchy, and log Student t distributions can be fitted by first taking logarithms of the response variable, as for the log normal distribution. The log logistic may be of special interest for durations because the intensity function is explicitly available (Section 4.5.1). Nevertheless, care must be taken because, as for the log normal distribution, the parameters of all these distributions generally change meaning. However, for the Laplace distribution, μ will still be the median. (Why?)

It is rarer to take logarithms when a distribution is already skewed, although this is possible with any of the duration distributions of Section 4.5. However, one well known distribution can be obtained in this way.

Pareto distribution   Thus, a new distribution can be obtained from the exponential distribution by taking logarithms of the response variable. However, this is usually generalised by introducing another parameter, say δ. The result is the Pareto distribution

$$\begin{aligned} \pi_i &= f(y_i; \lambda, \delta)\Delta_i \\ &= \lambda\delta^\lambda y_i^{-\lambda-1}\Delta_i \qquad \lambda > 0, \quad y_i \geq \delta > 0 \end{aligned}$$

(Note that here λ is no longer the constant intensity.) This model was originally developed to describe the distribution of high revenues, the upper tail of some more complex distribution. It is the continuous analogue of the zeta distribution.

The maximum likelihood estimates are

$$\hat{\delta} = \min(y_i)$$

$$\hat{\lambda} = \frac{n_\bullet}{\sum_i n_i \log(y_i/\hat{\delta})}$$

The theoretical mean and variance are

$$\mu_T = \frac{\lambda\delta}{\lambda - 1}, \qquad \lambda > 1$$

$$\sigma_T^2 = \frac{\lambda\delta^2}{(\lambda - 1)^2(\lambda - 2)}, \qquad \lambda > 2$$

An example is plotted in Figure 4.36. The curve decreases much faster than the exponential, but has a longer tail.

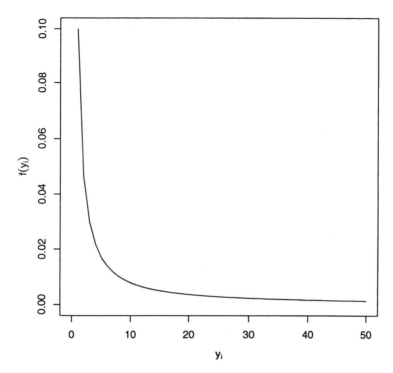

Fig. 4.36. The Pareto density function with $\delta = 1$ and $\lambda = 0.1$.

Example (continued)
Let us apply this model to the durations between coal-mining disasters for the first period, 1851–1891, given in Table 4.17. Table 4.22 reproduces these data and gives the fit of the Pareto distribution. The smallest observation (in the raw data) is $\hat{\delta} = 4$ and we also have $\hat{\lambda} = 0.350$. The curve is plotted in Figure 4.37. We see that this distribution overpredicts for both the shortest duration and the very long durations, as might be expected. This is also reflected in the AIC of 73.5 as compared with 5.4 for the exponential and 13 for the multinomial.   □

4.6.2   EXPONENTIAL TRANSFORMATION

If a new distribution is obtained by a log transformation of the response, going in the reverse direction requires an exponential transformation. Here, after substitution, the density must be multiplied by e^{y_i}.

Example
A normal distribution can be obtained from the log normal distribution by the transformation e^{y_i}.   □

Table 4.22. Days from Table 4.16, for the period 1851–1891, with the fitted Pareto distribution and the residuals.

Time	Disasters	Multinomial	Pareto	Residual
0–20	22	0.177	0.318	0.714
20–40	17	0.137	0.115	0.246
40–60	13	0.105	0.058	−0.155
60–80	8	0.065	0.037	−1.032
80–100	14	0.113	0.026	1.363
100–120	6	0.048	0.020	−0.786
120–140	9	0.073	0.016	0.756
140–200	12	0.097	0.033	−0.793
200–260	11	0.089	0.022	0.602
260–320	4	0.032	0.016	−0.670
320–380	4	0.032	0.013	0.326
380–440	2	0.016	0.010	−0.047
440–500	0	0.000	0.008	−1.122
> 500	2	0.016	0.308	−0.288

This transformation is less common but one important case needs to be discussed.

Extreme value distribution If we replace y_i by e^{y_i} in the Weibull distribution (Section 4.5.3), we obtain an extreme value distribution,

$$\pi_i = f(y_i; \alpha, \mu)\Delta_i$$
$$= \frac{\alpha e^{y_i \alpha} e^{-(e^{y_i}/\mu)^\alpha}}{\mu^\alpha} \Delta_i, \qquad \mu, \alpha > 0$$

As its name suggests, this is another distribution which, like the Pareto distribution, can describe extreme phenomena, such as the distribution of the age of the oldest person in each village, the period of longest illness in each hospital, or the height of the largest flood each year.

This distribution can be estimated by transforming the data to $z_i = e^{y_i}$ and applying the techniques for the Weibull distribution.

4.6.3 POWER TRANSFORMATIONS

The preceding transformations use known functions of the response. However, it is also possible that the transformation contain an unknown parameter to be estimated. The most common is a power transformation: y_i^α. When this is substituted into a density function, the function must be multiplied by $\alpha y_i^{\alpha-1}$.

Weibull distribution The Weibull distribution (Section 4.5.3) can be produced as a power transformation of the response in an exponential distribution.

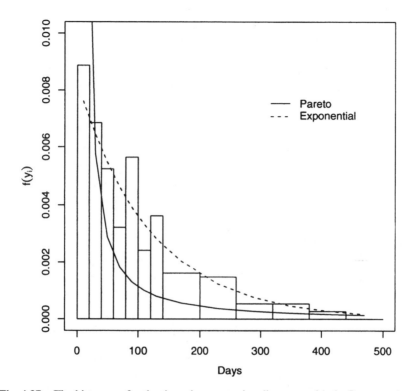

Fig. 4.37. The histogram for the times between mine disasters, with the Pareto and exponential density functions.

Power-transformed normal response   Traditionally, response variables have often been transformed to normality in order to be able to use the classical normal distribution models which we shall study in Chapter 5. The most common way is to estimate a power transformation. In this context, it is called the *Box–Cox transformation*. Unfortunately, the resulting expression is not a proper distribution because y_i^α must be positive (except for very special values of α) whereas the normal distribution requires that the response take any value. (The problem does not arise with the log transformation because logarithms can be negative.)

This transformation yields

$$f(y_i; \mu, \sigma^2, \alpha) = \frac{\alpha y_i^{\alpha-1}}{\sqrt{2\pi\sigma^2}} e^{-\frac{1}{2\sigma^2}(y_i^\alpha - \mu)^2}$$

This curve is asymmetric or skewed, except for $\alpha = 1$, with a longer tail to the right for $\alpha < 1$, and to the left for $\alpha > 1$. Where possible, the idea is to look for some interpretable transformation, such as the reciprocal ($\alpha = -1$) or logarithm ($\alpha \to 0$). Estimation of the parameters is really only feasible with a computer.

However, the results are generally even more difficult to interpret in terms of the original observations than the other transformations presented above.

4.7 Special families

Throughout the history of statistics, many of the available probability distributions have been classified into families in a number of different ways. In the modelling context, two of these families are of special interest.

4.7.1 LOCATION–SCALE FAMILY

Members of the location–scale family have two distinct parameters, for the location μ and the scale σ, that can vary independently. The response variable only appears in the density as $(y_i - \mu)/\sigma$. Thus, as we saw for the normal distribution (Section 4.4.1), the response can be transformed to a *standard deviate*:

$$z_i = \frac{y_i - \mu}{\sigma}$$

centred around zero with unit dispersion. The members of this family covered above (in Section 4.4) are the normal, logistic, Laplace, Cauchy, and Student t distributions. As we now know, within this family, μ and σ do not necessarily refer to the mean and standard deviation!

In contrast to the examples presented in this chapter, most modelling situations are more complex. As we saw in Chapter 2, usually the distribution of responses depends on the values of some explanatory variables. If the value of the location parameter μ depends in some way on explanatory variables, we have a regression model (Sections 2.2.5 and 2.3.3). The location–scale family is especially easy to generalise in this way because the shape of the density remains constant when the location parameter changes (for fixed dispersion parameter). As we have seen, this is often appropriate when only measurement errors are involved.

4.7.2 EXPONENTIAL FAMILY

Several of the distributions which we have studied in this chapter belong to another family, called the *exponential family*. However, their membership is not as obvious as for the location–scale family. Here, we must verify that their density can be written in the form

$$f(y_i; \theta) = \exp[y_i\theta + a(\theta) + b(y_i)] \tag{4.10}$$

where θ is some location parameter that is a function of the mean. It is called the *canonical parameter*. Notice that $b(\cdot)$ is a function of y_i only, whereas $a(\cdot)$ is a function only of θ.

When checking whether or not some distribution is a member of this family, a useful first step is to take the logarithm of the density function. Two members of this family covered in this chapter were the Poisson and binomial distributions.

Example

Consider the binomial distribution of Equation (1.9). This can be rewritten

$$\log[f(y_i; \pi_1)] = \log\left[\binom{n_\bullet}{y_i}\right] + y_i \log\left[\frac{\pi_1}{1 - \pi_1}\right] + n_\bullet \log(1 - \pi_1)$$

If we compare the terms with those in Equation (4.10), we see that

$$\theta = \log\left[\frac{\pi_1}{1 - \pi_1}\right]$$

$$a(\theta) = n_\bullet \log(1 - \pi_1)$$

$$= -n_\bullet \log\left[1 + e^\theta\right]$$

$$b(y_i) = \log\left[\binom{n_\bullet}{y_i}\right]$$

Thus, we discover that the canonical parameter is the logit, upon which logistic models are based. □

One major advantage of thinking in terms of such a general family is that many results are applicable in the same way to all members. This saves trying to figure them out separately for every model.

Sample size calculation A general formula can be written down for the calculation of sample size $n_\bullet$ for any member of the exponential family. From the logarithm of Equation (4.10), for $n_\bullet$ observations,

$$-\log[R(\theta)] = -\sum_i [y_i\theta + a(\theta) - y_i\hat{\theta} - a(\hat{\theta})]$$

$$= -n_\bullet[\bar{y}_\bullet(\theta - \hat{\theta}) + a(\theta) - a(\hat{\theta})]$$

Suppose that we want to detect a difference between the fixed values θ_0 and θ_1, where the latter is larger. Then, as we saw in Section 3.6, in the worst case, the likelihoods will be equal for these two parameter values. Thus, again using the logarithm of Equation (4.10), we shall have

$$y_\bullet\theta_0 + n_\bullet a(\theta_0) + \sum_i b(y_i) = y_\bullet\theta_1 + n_\bullet a(\theta_1) + \sum_i b(y_i)$$

$$\bar{y}_\bullet\theta_0 + a(\theta_0) = \bar{y}_\bullet\theta_1 + a(\theta_1)$$

or

$$\bar{y}_\bullet = \frac{a(\theta_0) - a(\theta_1)}{\theta_1 - \theta_0}$$

Thus, because we know the values of θ_1 and θ_0, we can calculate $\bar{y}_\bullet$ for this situation. Then, because θ is a known function of the mean, we can also obtain $\hat{\theta}$. Finally, we find the required sample size to be

$$n_\bullet = \frac{-\log(R)}{\bar{y}_\bullet(\hat{\theta} - \theta_0) - a(\theta_0) + a(\hat{\theta})}$$

where $-\log(R)$ is the fixed size of negative log normed likelihood to be used in making inferences.

Example
For the binomial distribution, we may want to know how many Bernoulli trials will be required to distinguish between, say, π_{10} and π_{11}. We have

$$\begin{aligned} \bar{y}_\bullet &= \hat{\pi}_1 \\ &= \frac{\log(1 + e^{\theta_1}) - \log(1 + e^{\theta_0})}{\theta_1 - \theta_0} \\ &= \frac{\log(1 - \pi_{10}) - \log(1 - \pi_{11})}{\log\left(\frac{\pi_{11}}{1 - \pi_{11}}\right) - \log\left(\frac{\pi_{10}}{1 - \pi_{10}}\right)} \end{aligned}$$

and

$$n_\bullet = \frac{-\log(R)}{\bar{y}_\bullet\left[\log\left(\frac{\bar{y}_\bullet}{1 - \bar{y}_\bullet}\right) - \log\left(\frac{\pi_{10}}{1 - \pi_{10}}\right)\right] - \log(1 - \pi_{10}) + \log(1 - \bar{y}_\bullet)}$$

If we want to distinguish between $\pi_{10} = 0.6$ and $\pi_{11} = 0.7$, we find that $\bar{y}_\bullet = 0.65$. If we are going to use the AIC, $-\log(R) = 1$ for a difference of one estimated parameter, so that $n_\bullet = 188.97$ or 189 Bernoulli trials are required.   □

Exponential dispersion family   The exponential family can be further generalised by introducing a dispersion parameter σ^2 (such as the variance). This yields the *exponential dispersion family* with

$$f(y_i; \theta, \sigma^2) = \exp\left[\frac{y_i\theta + a(\theta)}{\sigma^2} + b(y_i, \sigma^2)\right]$$

In the exponential family above, the dispersion parameter is unity. Again, θ is the canonical parameter.

The Poisson, binomial, normal, gamma, and inverse Gauss distributions studied in this chapter are all members of this family. The latter three can also be written as two-parameter members of the exponential family.

Example
Let us look at a two-parameter distribution, such as the normal distribution. We can write this density function as

$$\log[f(y_i; \mu, \sigma^2)] = -\frac{1}{2}\log(2\pi\sigma^2) - \frac{1}{2\sigma^2}(y_i - \mu)^2$$

$$= -\frac{1}{2}\log(2\pi\sigma^2) - \frac{y_i^2}{2\sigma^2} + \frac{y_i\mu - \mu^2/2}{\sigma^2}$$

so that

$$\theta = \mu$$

$$a(\theta) = -\frac{\mu^2}{2}$$

$$= -\frac{\theta^2}{2}$$

$$b(y_i, \sigma^2) = -\frac{1}{2}\log(2\pi\sigma^2) - \frac{y_i^2}{2\sigma^2}$$

Here, the mean is the canonical parameter.   □

As with the location–scale family, the location parameter of either of these families can also be allowed to vary with explanatory variables. When this is through a linear function, we have a *generalised linear model*. Notice however that, in contrast to regression models using the location–scale family, here the shape does change when the location parameter varies, except for the normal distribution.

All the logistic and log linear models of Chapter 2 were members of the generalised linear model family. In the next chapter, I shall look at other examples: models based on the normal distribution. These will be members of both the location–scale and generalised linear model families.

4.8 Exercises

(1) In the text, I fitted a uniform distribution to the numbers of ill children born each month, given in Table 4.1. My model did not take into account the fact that months have different numbers of days.

    (a) Construct a new model, based on the uniform distribution, using this information.

    (b) Does it fit better than the previous model?

(2) The table on the following page gives the counts of occurrence of surnames from a study area of Reading, Workingham, and Henley-on-Thames, England (Fox and Lasker, 1983). The names for the complete study were those of all 2397 couples whose marriages were registered in the study area during a 12-month period in 1972–1973. Those given in the table are for one of eight districts of that area. Although the sample may have been reasonably representative of surnames in that geographical area, it is clearly not random with respect to age or other characteristics. As well, some of the people may only have been in the area for the purpose of registering their marriage.

Number of occurrences	Number of surnames
1	329
2	43
3	11
4	1
5	0
6	1
7	0
8	0
9	1

(a) Choose an appropriate probability distribution and fit it.

(b) Calculate the AIC and check the residuals.

(c) Discuss how well the model fits.

(d) What general conclusions could be drawn, given the way in which the sample was selected?

(3) The table in Exercise (2.10) gave the frequencies of different numbers of burglaries in Detroit, USA.

(a) Choose an appropriate probability distribution and explain your choice.

(b) Fit the distribution.

(c) Calculate the AIC and check the residuals.

(d) Discuss how well the model fits.

(4) Let us reconsider the accident data in the two tables of Exercise (1.7).

(a) Choose appropriate probability distributions for each, explaining why.

(b) Fit them.

(c) Calculate the AICs and check the residuals.

(d) Discuss how well the models fit to each table.

(5) Table 2.22 gave the counts of accidents to men working in a soap factory over a five month period.

(a) Choose an appropriate probability distribution, giving your reasons.

(b) Fit the distribution.

(c) Calculate the AIC and check the residuals.

(d) Discuss how well the model fits.

(6) The table on the next page gives the numbers of deaths by horse kicks in 10 Prussian army corps over a 20 year period, that is, for 200 corps–years (Sokal and Rohlf, 1969, p. 94).

Deaths	Corps
0	109
1	65
2	22
3	3
4	1

(a) What is a reasonable model to describe the way in which these deaths might have occurred?

(b) Fit the model and explain your conclusions.

(7) The table below gives the number of fire losses per year from 1950 to 1973 for the buildings in a major university (Aiuppa, 1988, from Cummins and Freifelder).

Number of losses	Number of years
0	0
1	3
2	7
3	2
4	5
5	1
6	3
7	2

(a) Choose an appropriate probability distribution and justify your choice.

(b) Fit the distribution.

(c) Calculate the AIC and check the residuals.

(d) Discuss how well the model fits.

(e) In fact, the number of buildings at the university evolved over the years concerned, from 273 in 1950 to 312 in 1962. If such data were available for each year, discuss how this could be taken into account.

(8) Let us reconsider the consumer purchasing data of Exercise (1.8).

(a) Choose an appropriate probability distribution and fit it.

(b) Calculate the AIC and check the residuals.

(c) Discuss how well the model fits.

(9) During a cholera epidemic in India, the number of cases in each house was recorded (Dahiya and Gross, 1973):

Cases	Houses
0	168
1	32
2	16
3	6
4	1

(a) Fit an appropriate distribution to these data and draw your conclusions.

(b) Some of the houses which registered no cases were probably already infected. Fit a model without using this category and use it to predict the number of such houses among the 168.

(10) In Table 4.8, I studied the distribution by sex in families of 12 children in Saxony. The following table (Fisher, 1958, p. 67) gives the results from the same study for families of 8 children.

Males	Families
0	215
1	1485
2	5331
3	10649
4	14959
5	11929
6	6678
7	2092
8	342

(a) Is a binomial distribution suitable in this case?

(b) Do the residuals indicate the same sort of departures from this model as for families of 12 children?

(c) What conclusions can be drawn from the analysis of the two data sets?

(11) The number of children ever born to a sample of mothers over 40 years of age was collected by the East African Medical Survey in the Kwimba district of Tanganyika (Brass, 1959):

Children	1	2	3	4	5	6	7	8	9	10	11	12
Mothers	49	56	73	41	43	23	18	18	7	7	3	2

(a) List the ways in which these data are different than those in Table 4.8 and in Exercise (4.10)

(b) Try to fit a suitable distribution for these data. (A somewhat similar data set, the postal survey, was given as an example in Section 2.3.3. Recall also the discussion of truncated distributions in Section 4.2.2.)

(12) For the divorce data of Table 4.18,

(a) How well does the inverse Gauss distribution fit to these data as compared with those tried in the text?

(b) What is a possible interpretation of this model in this context?

(c) Does any transformation of the data yield a reasonable model for these data?

(13) The table on the next page shows the duration of strikes in the UK that began in 1965, as recorded by the Ministry of Labour (Lancaster, 1972).

Those given lasted more than one day and involved at least 10 people; they are for metal manufacturing and for all industries except transport and electrical machinery. One day strikes have not been included because they often are of a different nature, being a token stoppage appearing as a demonstration or threat. In the period considered, the majority of strikes in the UK were not claims for wage increases, but about questions of discipline, hours of work, sympathy, union recognition, and so on.

Duration	Number of strikes Metal	All	Duration	Number of strikes Metal	All
2	43	203	10	3	23
3	37	149	11–15	16	61
4	21	100	16–20	4	27
5	19	71	21–25	4	17
6	11	49	26–30	3	16
7	8	33	31–40	3	16
8	8	29	41–50	5	12
9	9	26	> 50	4	8

(a) Choose appropriate probability distributions and fit them to each data set.
(b) Compare the results.
(c) Calculate the AICs and check the residuals.
(d) Discuss how well the models fit to each and what can be learnt about the process by which a strike comes to a conclusion.

(14) A survey was made of women having a bachelor's but no higher degree and employed as mathematicians or statisticians. Monthly salaries (dollars) of these female mathematics graduates involved in nonsupervisory positions are given below (Zelterman, 1987, from Beatty).

Monthly salary	No.	Monthly salary	No.	Monthly salary	No.
1051–1150	1	2151–2250	11	3251–3350	1
1151–1250	1	2251–2350	6	3351–3450	4
1251–1350	6	2351–2450	11	3451–3550	1
1351–1450	3	2451–2550	3	3551–3650	2
1451–1550	4	2551–2650	4	3651–3750	0
1551–1650	3	2651–2750	5	3751–3850	2
1651–1750	9	2751–2850	6	3851–3950	0
1751–1850	6	2851–2950	4	3951–4050	0
1851–1950	5	2951–3050	4	4051–4150	1
1951–2050	16	3051–3150	5		
2051–2150	4	3151–3250	1		

(a) Choose an appropriate probability distribution and fit it.
(b) Calculate the AIC and check the residuals.

(c) Discuss how well the model fits.

(15) The following table gives the number of years since their first degree of the same sample of female mathematics graduates practising mathematics or statistics as described in Exercise (4.14) above. (Zelterman, 1987).

Years	Number	Years	Number	Years	Number
0	5	10	2	22–23	3
1	14	11	3	24–25	4
2	10	12	3	26–27	3
3	8	13	3	28–29	1
4	11	14	0	30–31	1
5	4	15	1	32–33	2
6	3	16	5	34–35	0
7	5	17	2	36–40	6
8	7	18–19			
9	5	20–21			

(a) Choose an appropriate probability distribution and fit it.

(b) Calculate the AIC and check the residuals.

(c) Discuss how well the model fits.

(16) Let us look again at the event recall data of Exercise (1.6).

(a) What might be an appropriate probability distribution for these data?

(b) Fit the model.

(c) Calculate the AIC and check the residuals.

(d) Discuss how well the model fits.

(17) Employment durations of recruits to the British Post Office in the first quarter of 1973 (Burridge, 1981) were given in Table 4.6. In fact, there were two groups corresponding to different grades, as shown in the following table.

Quarters	Group A	B	Quarters	Group A	B	Quarters	Group A	B
1	22	30	9	2	3	17	1	1
2	18	28	10	1	0	18	1	0
3	19	31	11	0	0	19	3	2
4	13	14	12	1	1	20	1	0
5	5	10	13	0	1	21	1	3
6	6	6	14	0	0	22	0	1
7	3	5	15	0	0	23	0	1
8	2	2	16	1	1	24	0	0

(a) Choose and fit a suitable probability distribution to each group.

(b) Compare the results and discuss how well each model fits.

(c) Have you found a continuous distribution that fits better than the geometric distribution?

- (d) Calculate the AICs and check the residuals.
- (e) Combine the data for the two groups and refit the model.
- (f) Does it change very much from the two separate models?
- (g) Because observations on the two groups are independent, the AICs for the two groups separately can be added together and compared with that for the model where the groups were combined. What can be said about the difference between the groups?

(18) The inverse Gauss and log normal distributions are closely related to the normal distribution. The latter is in the location–scale family. Are the first two members of this family? Why?

(19) For each distribution studied in this chapter,
- (a) check whether it is a member of the exponential family,
- (b) a member of the exponential dispersion family, and
- (c) for the members of the exponential family, derive the sample size formula.

5
Normal regression and ANOVA

In Chapter 2, we explored models that allowed the probabilities of the various possible response categories to vary depending on the values of explanatory variables. However, we imposed no specific constraints as to the shape of the probability distributions under this different conditions (except for order in Section 2.3.4) so that a large number of parameters was often involved. In Chapter 4, we looked at how the forms of probability distributions could be simplified by adopting some appropriate density function. Two major advantages of this procedure were that an appropriate density function should be informative about the data generating mechanism and that the number of parameters to estimate was reduced.

It is now time to combine these two approaches. In principle, we can select any distribution from Chapter 4 and allow any parameter in it to vary depending on some explanatory variables. However, this is too advanced a topic for this text. Instead, I shall concentrate on one particular case, when the mean of a normal distribution (Section 4.4.1) depends on explanatory variables with constant variance.

5.1 General regression models

In Section 4.7, I mentioned two possible types of general regression models, based on the location–scale and exponential families. It will be useful to look at these again briefly here.

5.1.1 MORE ASSUMPTIONS OR MORE DATA

As we have now seen, the probability distribution for a response variable can take a wide variety of possible forms, depending on the underlying data generating mechanism. In very simple situations, as in Chapter 4, we can assume that the population is homogeneous so that the same distribution applies to all individuals. In most cases, however, we shall be interested in subgroups within which the response distribution may take different forms. In other words, the distribution of the response depends on one or more explanatory variables.

Examples (continued)
In Section 4.3.1, the distribution of years behind in Bombay primary schools was not the same in all social classes and, in Section 4.5.2, the distribution of times between mine disasters was not identical over the whole period of observation. □

If the same data generating mechanism were operating for all individuals, we might make the assumption that the same functional form for the distribution holds for all subgroups, with only the parameter values changing.

Example (continued)
This assumption was not valid for the Bombay data where only the years behind in primary school for children of certain social classes appeared to follow a Poisson distribution.   □

In most simple situations, we make a further simplifying assumption that only one parameter, the location, of the distribution varies among the subpopulations.

Example
One of the parameters in many density functions is the mean, which is usually easily interpretable. This mean may change under different circumstances. Thus, the mean income may vary by sex and by region of residence. Simple models, similar to those that we fitted in Chapter 2, may be set up to describe those differences.□

On the other hand, all response variables are empirically observed as discrete categories, defined by the unit of measurement. Thus, they can always be described by the multinomial distribution. The polytomous logistic categorical data models of Chapter 2 allow changes among subpopulations and could theoretically be applied to any response. If categorical data models are so flexible and so many forms of probability distribution are possible, one may then ask: why not use logistic or log linear models in all applications where explanatory variables influence the response?

We know, however, that such models have large numbers of parameters. To be able to estimate them, sufficient observations have to be available. The shape of the multinomial distribution, that is, of the histogram, has to be discernible. The data must define the shape, because no theoretical functional form of density is being imposed, in contrast to the models in Chapter 4.

As we saw in Section 4.1.2, when the response is a quantitative variable, it is usually preferable to work with some density function for several reasons:

(1) the density function smooths the histogram, making clearer what form it might have for the whole population;

(2) a simpler model is produced, with fewer parameters to estimate;

(3) fewer observations will be required in order to estimate the parameters reliably;

(4) such a simpler model may have more chance of being directly applicable to other similar studies;

(5) the density will hopefully provide information about the underlying data generating mechanism.

The basic trade-off is between more theoretical assumptions and more data. Models based on density functions reduce the number of parameters and the amount of data required to estimate them. In addition, if we want to learn about a data generating mechanism, we have to make theoretical assumptions about it and confront them with some data!

In models based on density functions, changes in the location parameter will determine changes in all the probabilities of different response values. A complex system of equations, one for each response category (Section 2.3), is no longer required.

5.1.2 GENERALISED LINEAR MODELS

In simple situations, changes in distribution can be modelled by having only one parameter, the mean or location parameter, change. This contrasts with polytomous logistic models where we have to model separately how each bar of the histogram varies among subpopulations (with the constraint that the probabilities sum to one). As in Chapter 2, we can use a linear function for the explanatory variables, for example,

$$g(\mu_i) = \beta_0 + \beta_1 x_i$$

where $g(\cdot)$ is some function. In Chapter 2, it was the logit or log function.

Example
Recall that the theoretical mean of a binomial distribution is $\mu = n_\bullet \pi_1$ (Section 1.3.4). Thus, a logit can be rewritten as a function of the mean:

$$\log\left(\frac{\pi_1}{1-\pi_1}\right) = \log\left(\frac{\mu/n_\bullet}{1-\mu/n_\bullet}\right)$$

$$= \log\left(\frac{\mu}{n_\bullet - \mu}\right)$$

We have also seen (Section 4.7.2) that the logit is the canonical parameter for this member of the exponential family. □

In a model based on some density function, the linear function will describe changes in the location parameter, not directly changes in individual conditional probabilities. This $g(\cdot)$, connecting the location parameter of the density to the linear function, is called a *link function*. As we have seen (Section 2.2.2), one important role of such functions is to ensure that the parameters of the probability distribution do not take impossible values.

With an appropriate choice of density, this is the family of *generalised linear models*. The simplest models in this family equate the linear function to the canonical parameter of a member of the exponential (dispersion) family. Besides the logistic and log linear models of Chapter 2, the only other member that I shall consider in this text will be based on the normal distribution, the subject of this chapter. It is particularly simple, because the canonical parameter is the mean. Thus, we can construct our models directly without any intervening link function (sometimes called the identity link function).

5.1.3   LOCATION REGRESSION MODELS

The location–scale family differs from most members of the exponential family in a number of ways that are relevant for regression modelling:

- the density is symmetric;
- the location parameter can take any value, positive or negative;
- the location parameter can change without altering the shape of the distribution (with fixed scale parameter).

These characteristics imply that, for this family, we generally shall not require a link function in the regression model, so that we can use

$$\mu_i = \beta_0 + \beta_1 x_i$$

The models based on the normal distribution that I shall look at in this chapter are simultaneously members of both these families.

5.2   Linear regression

In Sections 2.2.5 and 2.3.3, we encountered two examples of regression models: logistic and log linear regression. For reasons that we studied in Section 4.4.1, regression models based on the normal distribution will be simpler to calculate. The maximum likelihood estimates of the parameters can be obtained by 'least squares'.

Linear regression with a normal distribution is, perhaps, the most well known generalised linear model. With this model, we attempt to describe how a histogram having the form of a normal distribution changes, depending on explanatory variables, in different subgroups of the population. Models based on the normal distribution can be particularly simple because, as we have seen, the mean of this distribution can change without affecting the other parameter, the variance. Thus, such models can assume a constant variability in all segments of the population, an hypothesis that may or may not be reasonable. This is in addition to the assumption about the symmetric 'bell-shaped' form of the response distribution within each subgroup of the population.

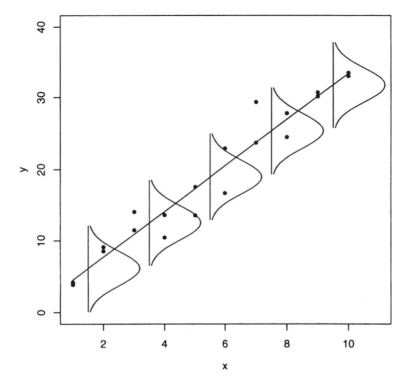

Fig. 5.1. Graphical representation of a simple linear normal regression. The vertical axis represents both y_i and its mean, μ_i. The axis coming out of the paper represents the probability of a given combination of y_i and x_i.

5.2.1 ONE EXPLANATORY VARIABLE

Let us first investigate the simple case where changes in the mean of the normal distribution depend only on one quantitative explanatory variable x_i in the following simple way:

$$\mu_i = \beta_0 + \beta_1 x_i \qquad (5.1)$$

The mean response is related to the explanatory variable by means of a straight line. β_0 is called the *intercept*; it gives the value of the mean when $x_i = 0$ (if this value of x_i is possible). β_1 is the *slope* of the regression line. If it is positive, the mean increases with increasing x_i; if it is negative, the mean decreases.

This is called a *simple linear regression model*. It is represented in Figure 5.1. We see how the *mean* responses μ_i are following a straight line as x_i changes, whereas the *actual* response values y_i are scattered about them. In this way, the location of the normal distribution curve is being displaced, but the curve still keeps the same shape, defined by the functional form of the normal distribution and by the constant variance.

Example
People's incomes may depend on the amount of education they have, measured in some way, such as the number of years. Then, the average income might increase with the number of years of education, in a way similar to the straight line in Figure 5.1. However, all people with the same education will not have the same income. They will be distributed around that mean, perhaps in the form of a normal distribution, as also illustrated in that graph. In fact, the distribution of income will be skewed, with a few very large values, so it would be likely to be better represented by a log normal rather than normal distribution.   □

One interesting thing about such a model is that the two parameters, β_0 and β_1, defining the line, plus the constant variance, σ^2, can be estimated even if we have far too few observations actually to study the form of the distribution at each value of x_i. This results from our strong theoretical assumptions about the shape of the distribution and about how its location changes with x_i. Such model estimation with so few observations was not possible for analogous models using the multinomial distribution in Chapter 2 because they made no assumptions about the shape of the distribution. There, we had to have enough observations to make a histogram at each value of x_i. But, on the other hand, here we shall not usually be able to check with the data the validity of these theoretical assumptions concerning the normal distribution.

Maximum likelihood estimates   The maximum likelihood estimates of the three parameters are given by

$$\widehat{\beta}_1 = \frac{\sum_i (x_i - \bar{x}_{\bullet})(y_i - \bar{y}_{\bullet})}{\sum_i (x_i - \bar{x}_{\bullet})^2}$$

$$= \frac{\sum_i x_i y_i - n_{\bullet} \bar{x}_{\bullet} \bar{y}_{\bullet}}{\sum_i x_i^2 - n_{\bullet} \bar{x}_{\bullet}^2}$$

$$\widehat{\beta}_0 = \bar{y}_{\bullet} - \widehat{\beta}_1 \bar{x}_{\bullet}$$

$$\widehat{\sigma}^2 = \frac{1}{n_{\bullet}} \sum_i (y_i - \widehat{\beta}_0 - \widehat{\beta}_1 x_i)^2$$

$$= \frac{1}{n_{\bullet}} \sum_i y_i^2 + \widehat{\beta}_0^{\,2} + \widehat{\beta}_1^{\,2} \frac{1}{n_{\bullet}} \sum_i x_i^2 - 2\widehat{\beta}_0 \bar{y}_{\bullet} - 2\widehat{\beta}_1 \frac{1}{n_{\bullet}} \sum_i x_i y_i + 2\widehat{\beta}_0 \widehat{\beta}_1 \bar{x}_{\bullet}$$

The second formulæ for $\widehat{\beta}_1$ and $\widehat{\sigma}^2$ are easier to calculate, but can produce greater rounding errors. (Note that $\widehat{\sigma}^2$ must be positive.) The formulæ for estimating the two regression parameters are the same as those used for logistic regression in Section 2.2.5 and for log linear regression in Section 2.3.3, but, here, $z_i = y_i$ and $w_i = 1$. Because the weights are constant, the estimation is exact, with no iterations required.

In the same way as in Section 4.4.1, these estimates for the regression parameters are the values making the variance smallest, given the assumptions of the model. For this reason, they are sometimes called the 'least squares' estimates: they yield the smallest sum of squares, $\sum(y_i - \mu_i)^2$, here under the assumption that $\mu_i = \beta_0 + \beta_1 x_i$.

Normed likelihood function   We now require the likelihood function to compare two normal distribution models. From Equation (4.5), we can derive the normed likelihood function for $n_\bullet$ observations, assuming all of them to have distinct values of y_i ($n_i = 1$ for all i),

$$R(\mu, \sigma^2) = \prod_i \frac{\pi_i}{\tilde{\pi}_i}$$

$$= \frac{(\sigma^2)^{-n_\bullet/2} e^{-\frac{1}{2\sigma^2}\sum_i(y_i-\mu_i)^2}}{(\hat{\sigma}^2)^{-n_\bullet/2} e^{-\frac{1}{2\hat{\sigma}^2}\sum_i(y_i-\widehat{\mu_i})^2}}$$

$$= \left(\frac{\sigma^2}{\hat{\sigma}^2}\right)^{-\frac{n_\bullet}{2}} e^{-\frac{1}{2\sigma^2}\sum_i(y_i-\mu_i)^2 + \frac{n_\bullet}{2}}$$

using $\hat{\sigma}^2 = \sum_i(y_i - \widehat{\mu_i})^2/n_\bullet$. Now, for any fixed values of the mean, the maximum likelihood estimate of the variance is $\tilde{\sigma}^2 = \sum_i(y_i - \mu_i)^2/n_\bullet$. When we substitute this into the normed likelihood function, the last factor disappears and we obtain the normed *profile* likelihood function for the mean alone:

$$R(\mu, \tilde{\sigma}^2) = \left(\frac{\tilde{\sigma}^2}{\hat{\sigma}^2}\right)^{-\frac{n_\bullet}{2}}$$

$$= \left[\frac{\sum_i(y_i - \mu_i)^2}{\sum_i(y_i - \widehat{\mu_i})^2}\right]^{-\frac{n_\bullet}{2}} \tag{5.2}$$

This is a simple generalisation of Equation (4.7). Notice that, to compare models with different means, we compare their variances in the likelihood function.

In any given model, the variance remains constant as the explanatory variable changes. But, if we change models, this constant variance will be different. A model with a smaller estimate of the variance will be more likely for the observed data, although the extra parameters required (loss of degrees of freedom) must also be taken into account.

Then, one simple way to write the AIC is $n_\bullet \log(\tilde{\sigma}^2)/2 + p$, where p is the number of parameters estimated (including the variance). Note that, if the estimated variance is small enough, the AIC specified in this way can take negative values.

In the linear regression model, we have $\mu_i = \beta_0 + \beta_1 x_i$, the theoretical mean for each observation. Often, we shall be particularly interested in checking if the model where the mean does not depend on the explanatory variable is reasonable.

In that case, the slope $\beta_1 = 0$, so that $\mu_i = \beta_0$, with estimate, $\widehat{\beta}_0 = \bar{y}_\bullet$. The corresponding estimates of the variance for the two models are $\tilde{\sigma}^2 = \Sigma_i(y_i - \bar{y}_\bullet)^2/n_\bullet$ and $\hat{\sigma}^2 = \Sigma_i(y_i - \widehat{\beta}_0 - \widehat{\beta}_1 x_i)^2/n_\bullet$. The first can never be larger than the second. (Why?) Thus, from Equation (4.7), the deviance comparing these two models is

$$D(\beta_1 = 0) = n_\bullet \log \left(\frac{\Sigma_i(y_i - \bar{y}_\bullet)^2}{\Sigma_i(y_i - \widehat{\beta}_0 - \widehat{\beta}_1 x_i)^2} \right)$$

Because we have fixed one parameter, β_1, this has one degree of freedom.

Because the normal distribution is symmetric, this is one instance where standard errors represent the profile likelihood function very closely, as we have already seen in Section 4.4.1. For β_1, the parameter of interest, the standard error is

$$\frac{\sqrt{\Sigma_i(y_i - \widehat{\beta}_0 - \widehat{\beta}_1 x_i)^2/(n_\bullet - 2)}}{\sqrt{\Sigma_i(x_i - \bar{x}_\bullet)^2}}$$

from which likelihood intervals can be obtained, as in Section 3.2.4.

If a test of significance is required, the χ^2 distribution can be applied to the deviance. However, usually, an 'exact' test is used, calculated as

$$t_{n_\bullet - 2} = \frac{|\widehat{\beta}_1 - \beta_1|\sqrt{\Sigma_i(x_i - \bar{x}_\bullet)^2}}{\sqrt{\Sigma_i(y_i - \widehat{\beta}_0 - \widehat{\beta}_1 x_i)^2/(n_\bullet - 2)}} \tag{5.3}$$

for any value of β_1, although usually one looks at $\beta_1 = 0$. Notice that this formula is just the difference between two values of β_1 divided by the standard error.

This is called the *Student t statistic*. As might be expected from its name, this has a Student t distribution (Section 4.4.5). P-values can be obtained from Table A.2 of the Appendix or from computer software.

Example

Table 5.1 gives the ages and weights of 24 babies born in the same hospital. For the moment, I shall ignore their sex and study how weight depends on age alone. The parameters of a linear regression are estimated to be $\widehat{\beta}_0 = -1485.0$ and $\widehat{\beta}_1 = 115.5$, with constant variance $\hat{\sigma}^2 = 37094.29$. Thus, under this model, the weight increases on average by 115.5 g per week and has a standard deviation of 184.4 g. Notice, however, that the intercept cannot be interpreted as the mean weight at age zero ($x_i = 0$)!

This regression line is plotted in Figure 5.2. From the results in Section 4.4.1, we can calculate intervals around this line within which responses have some specified probability of occurring according to this model. For example, responses have a probability of 0.68 of being within one standard deviation of

Table 5.1. Weights (g) and ages (weeks) of babies. (Dobson, 1990, p. 17)

Male		Female	
Age	Weight	Age	Weight
40	2968	40	3317
38	2795	36	2729
40	3163	40	2935
35	2925	38	2754
36	2625	42	3210
37	2847	39	2817
41	3292	40	3126
40	3473	37	2539
37	2628	36	2412
38	3176	38	2991
40	3421	39	2875
38	2975	40	3231

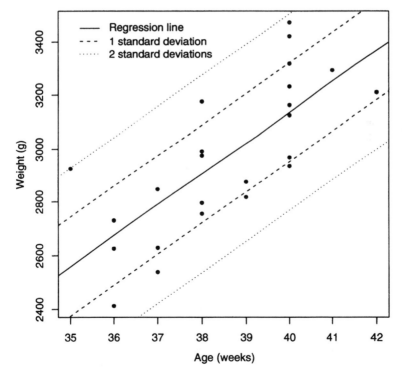

Fig. 5.2. Weights and ages of 24 babies, with the fitted normal linear regression line and the intervals of one and two standard deviations.

the mean, defined by the regression line, and 0.95 of being within two standard deviations. These intervals are also shown in Figure 5.2. They provide an idea of the form of the normal distribution, as was illustrated in Figure 5.1.

This regression model has an AIC of 128.2. For comparison, the model with constant weight for all ages has an AIC of 136.9, so that, as might be expected, there is strong evidence that weight changes with age.

The deviance for the constant weight model, as compared with the linear regression model, is 19.4 with one degree of freedom. If we wish to apply a significance test, we should not require a table do know that this deviance is significant!

The Student t value is 5.2 with 22 d.f. We can look this up in Table A.2. However, there, we have a choice of two significance levels. If possible models can lie on either side of our hypothesis, we can use a two-sided test, otherwise a one-sided test. In the first case, we do not know, *a priori*, what effect the explanatory variable might have, whereas, in the second, we know the direction of the effect. Here, we do not expect weight to be lost as age increases, so that β_1 should be greater than one. We can use a one-sided test and reach the same conclusions as before.

However, comparison of models or tests of null hypotheses are never sufficient. We require information about the precision of important parameters. This is provided by the normed profile likelihood for β_1 plotted in Figure 5.3. We can see that a fairly large range of values of the slope is plausible. From this graph, summary intervals can be obtained, if required, as in Section 3.4.   □

Statistical software usually only provides results for the null model with no difference: $\beta_1 = 0$. This is the most common case, but others are also possible.

Example (continued)
Table 1.7 and Figure 1.5 displayed data on the relationship between weight before and after diet. When we fit the linear regression model, we obtain $\widehat{\beta}_0 = 4.187$, $\widehat{\beta}_1 = 0.910$, and $\hat{\sigma}^2 = 4.190$, with AIC 10.2. The regression function for this model is plotted in Figure 5.4, along with the data. The estimated slope indicates that, on average, after diet the people are 91% of the weight before diet.

If the slope were one, one might think that people would, on average, have the same weight after as before diet. Closer examination shows that this is not correct. The regression equation then becomes

$$\mu_i = \beta_0 + x_i \qquad (5.4)$$

so that the average weight μ_i after diet has a constant difference β_0 from the prediet weight x_i. (Notice that such a construction is really only meaningful in the location-scale family.) Thus, for the weight to be the same before and after diet, we would also require $\beta_0 = 0$. Dependence of postdiet weight on prediet weight with $\beta_1 \neq 1$ is not the same as dependence of change in weight on diet!

Let us, nevertheless, examine the model for constant change in weight, $\beta_1 = 1$, $\beta_0 \neq 0$. This model has $\widehat{\beta}_0 = -1.8$ after diet. The variance for this model is

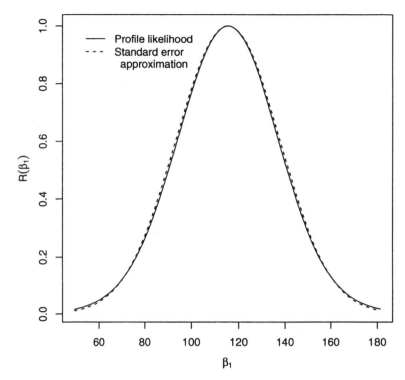

Fig. 5.3. Exact and approximate normed profile likelihoods for the slope parameter for dependence of weight on age.

estimated to be 4.76, giving an AIC of 9.8. Thus, there is no evidence against this latter model; however, there are only 10 observations. On the other hand, the hypothesis that the weight after diet does not depend on weight before, that is, that $\beta_1 = 0$, is not very plausible. Thus, as would be expected, the AIC is 45.4.

If required, we can also calculate the Student t values: that for $\beta_1 = 0$ is 10.60, whereas that for $\beta_1 = 1$ is 1.04, both with 8 d.f. Again, we have a choice of two significance levels. Here, we do not expect weight to be gained after such a diet, so that β_1 should not be greater than one. We can use a one-sided test and see that the test for $\beta_1 = 0$ is extremely significant, whereas that for $\beta_1 = 1$ does not even have a P-value of 0.10.

In Section 5.3.3, I shall examine the model with no change in weight, that is, $\beta_0 = 0$. ☐

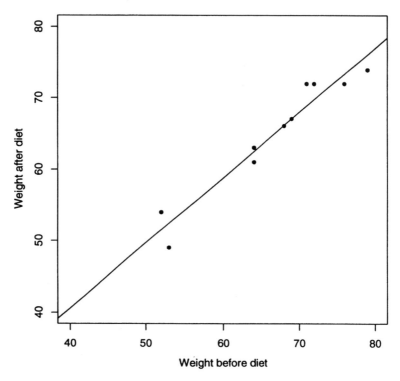

Fig. 5.4. Changes in weight (kg) with diet for ten people, with the fitted normal linear regression line.

5.2.2 MULTIPLE REGRESSION

The simple linear regression model can be directly extended to several explanatory variables, called multiple regression:

$$\mu_i = \beta_0 + \beta_1 x_{1i} + \beta_2 x_{2i} + \beta_3 x_{3i} + \cdots$$

However, the amount of calculation makes a computer necessary. As an illustration, for two explanatory variables, the parameter estimates are given by

$$\widehat{\beta}_1 = \frac{\sum_i x_{1i} y_i \sum_i x_{2i}^2 - \sum_i x_{2i} y_i \sum_i x_{1i} x_{2i}}{\sum_i x_{1i}^2 \sum_i x_{2i}^2 - (\sum_i x_{1i} x_{2i})^2}$$

$$\widehat{\beta}_2 = \frac{\sum_i x_{2i} y_i \sum_i x_{1i}^2 - \sum_i x_{1i} y_i \sum_i x_{1i} x_{2i}}{\sum_i x_{1i}^2 \sum_i x_{2i}^2 - (\sum_i x_{1i} x_{2i})^2}$$

$$\widehat{\beta}_0 = \bar{y}_\bullet - \widehat{\beta}_0 - \widehat{\beta}_1 \bar{x}_{1\bullet} - \widehat{\beta}_2 \bar{x}_{2\bullet}$$

$$\widehat{\sigma}^2 = \frac{1}{n_\bullet} \sum (y_i - \widehat{\beta}_0 - \widehat{\beta}_1 x_{1i} - \widehat{\beta}_2 x_{2i})^2$$

The Student t value is

$$t_{n_\bullet - 3} = \frac{|\widehat{\beta}_1 - \beta_1|\sqrt{\sum_i x_{1i}^2 \sum_i x_{2i}^2 - (\sum_i x_{1i}x_{2i})^2}}{\sqrt{\sum_i x_{2i}^2 \sum (y_i - \widehat{\beta}_0 - \widehat{\beta}_1 x_{1i} - \widehat{\beta}_2 x_{2i})^2 / (n_\bullet - 3)}}$$

and similarly for β_2. As above, this is the difference in β_j values divided by the standard error.

Example (continued)
For the weights of the babies, we also have available the sex as a second explanatory variable. Let us code this as 1 for males and 0 for females. Then, the regression coefficients are estimated to be $\widehat{\beta}_0 = -1773.3$, $\widehat{\beta}_1 = 120.9$ (for age), and $\widehat{\beta}_2 = 163.0$ (for sex), with variance $\widehat{\sigma}^2 = 27\,448.78$. Notice that the variance is considerably smaller and that the previous estimates of both the regression parameters have changed. The AIC is 126.6 as compared with 128.2 above so that this model is an improvement on that without sex.

According to this model, all babies increase in weight on average by 120.9 g per week. However, at all ages, the male babies weigh 163.0 g more on average. This can be seen clearly in Figure 5.5 where the regression function is plotted.

As with models involving two explanatory variables in Chapter 2, we can include an interaction between sex and age. This will allow the slope to be different for the two sexes. To do this here, we can multiply the two variables together to create a new one and include this in the model. However, the AIC is 127.5, showing that this interaction is not necessary. This means that the slopes for the two sexes can be assumed to be the same.

For information about the precision of the two regression parameters, the normed profile likelihoods for β_1 and β_2 are plotted in Figure 5.6. We can see that especially the parameter relating to sex differences is not estimated very precisely. Again, these graphs can be summarised by producing intervals, if required, as in Section 3.4.   □

Thus, multiple regression is a powerful tool, not restricted to quantitative variables. Indeed, as we saw in Section 2.2.5, the procedure used in this example can be generalised to handle nominal explanatory variables with more than two categories. Qualitative variables can be recoded as a set of indicator variables. This set will have one less new variable than the number of categories.

In Section 2.2.1, we saw two different ways of placing constraints on a categorical variable. The same situation applies here. If we want a baseline constraint, the set of new variables will correspond to all categories except the baseline. Each will have 1 for individuals having the given category and 0 otherwise. If we want a mean constraint, we must arbitrarily choose one category to exclude. (The choice will not affect the results.) The set of new variables will be the same as that for the baseline constraint except that individuals in the excluded category will have the code -1.

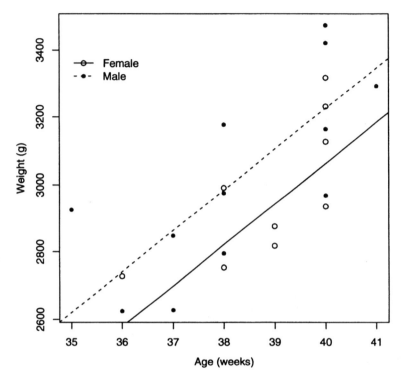

Fig. 5.5.   Weights and ages of 24 babies, with the fitted normal linear regression lines allowing for sex.

Example
Suppose that a variable has four categories, A, B, C, and D. With A as the baseline, the values in the three new indicator variables will be

	1	2	3
A	0	0	0
B	1	0	0
C	0	1	0
D	0	0	1

For the mean constraint, arbitrarily excluding D, the values will be

	1	2	3
A	1	0	0
B	0	1	0
C	0	0	1
D	−1	−1	−1

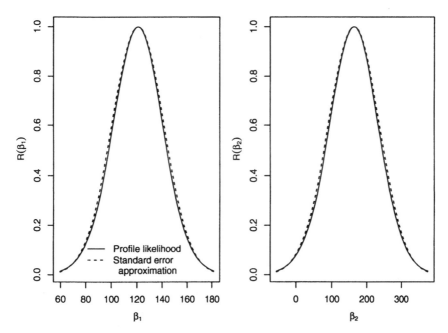

Fig. 5.6. Exact and approximate normed profile likelihoods for the slope parameters for dependence of weight on age and sex.

Most statistical software now handles the creation of such variables automatically in a user-friendly way. □

We saw in Sections 2.2.5 and 2.3.3 that linear regression models can also be applied for a binary response variable as logistic regression and for other frequency tables as log linear regression. The same type of linear regression approach can also be used for any other distribution of Chapter 4. However, suitable computer software is only widely available for the generalised linear model family.

5.3 Analysis of variance

As we have now seen, multiple regression can be used to handle all explanatory variables, quantitative or categorical, upon which the mean of a normal distribution depends. This technique is now generally available in computer software to handle all such variables. However, simpler methods of calculation, suitable for a hand calculator, exist when all explanatory variables are categorical. These models are analogous to the logistic and log linear models in Chapter 2. I shall briefly look at these here, continuing with the assumption that the responses follow a normal distribution and that their variance is the same in all subgroups.

5.3.1 ONE EXPLANATORY VARIABLE

As we have seen, when the explanatory variable is nominal, with several categories, one possibility is to create an indicator variable for each category. If there are J categories, only $J-1$ of these indicator variables is necessary. Then, a multiple regression can be applied as above. However, this involves a fair bit of calculation.

Another technique, which gives exactly the same results, is to construct a model similar to those in Chapter 2:

$$\mu_j = \mu + \alpha_j \tag{5.5}$$

again, with the appropriate constraints, summing to zero or with a baseline category. Suppose that there are n_j observations y_{ij} in category j. The maximum likelihood estimates are $\widehat{\mu_j} = \bar{y}_{\bullet j} = \sum_i y_{ij}/n_j$. These can be substituted into the left-hand side of Equation (5.5), the solution giving the maximum likelihood estimates of μ and α_j. The procedure is the same as that in Chapter 2 (only now no logarithms are needed!).

The likelihood or deviance can still be obtained from Equations (4.7) and (5.2). Again a significance test can be based on this, but the classical way of presenting the results is slightly more complex. We have seen that likelihood comparisons of normal models with different means involve comparison of their respective variances. This has given the traditional name to these models: *analysis of variance* or ANOVA. A summary table can be set up:

	SS	MSS	d.f.	F
Total effect	$S_1 = \sum_{ij}(y_{ij} - \bar{y}_{\bullet\bullet})^2$	$\frac{S_1}{n_\bullet - 1}$	$n_\bullet - 1$	
Main effect	$S_2 = \sum_j n_j(\bar{y}_{\bullet j} - \bar{y}_{\bullet\bullet})^2$	$\frac{S_2}{J-1}$	$J-1$	$\frac{S_2/(J-1)}{S_3/(n_\bullet - J)}$
Residual	$S_3 = \sum_{ij}(y_{ij} - \bar{y}_{\bullet j})^2$	$\frac{S_3}{n_\bullet - J}$	$n_\bullet - J$	

where SS stands for 'sum of squares' and MSS for 'mean sum of squares'. The last line is called the residual because it is the sum of the squares of the residuals for this model.

The value in the last column is the *F statistic*, with $J-1$ and $n_\bullet - J$ degrees of freedom, which can be used for a significance test. Values from this F distribution for selected degrees of freedom are tabulated in Table A.3 of the Appendix. The values on the first page, for one numerator degree of freedom, are just the squares of the values for the Student t distribution in Table A.2.

An important special case of this one-way analysis of variance occurs when the explanatory variable has only two categories. This models the difference between two means. Equation (5.3) can, then, be rewritten as

$$t_{n_\bullet - 2} = \frac{|\widehat{\mu_1} - \widehat{\mu_2}|}{\sqrt{\left(\frac{1}{n_1} + \frac{1}{n_2}\right)\frac{1}{n_\bullet - 2}\sum_i \sum_j (y_{ij} - \bar{y}_{\bullet j})^2}}$$

Table 5.2. Times (hours) reported spent doing a homework assignment by the first class taking this course in 1976, classified by sex.

Male									
20	20	20	8	12	14	40	13	16	12.5
16	34	12	18	17	30				

Female									
16	13	22	14	12	6	5	25	24	15

Table 5.3. Analysis of variance for the homework data in Table 5.2.

	SS	MSS	d.f.	F
Total effect	1645.2	65.8	25	
Sex effect	84.5	84.5	1	1.30
Residual	1560.7	65.0	24	

to yield a significance test for two means being identical.

As always for inferences about scientific problems, summaries of selection criteria, or significance tests in analysis of variance tables, are not sufficient. The estimates and precision of all important parameters should be reported using the likelihood function.

Example (continued)
In Section 4.4, we looked at the times spent by students doing a homework project. The actual times, classified by sex, are given in Table 5.2. We shall here be interested in determining whether there is a difference in average time taken by the two sexes. The mean times are estimated to be 18.91 and 15.20, respectively. These give $\hat{\mu} = 17.05$ and $\widehat{\alpha_1} = 1.85 = -\widehat{\alpha_2}$ in Equation (5.5) using the mean constraint. Note that this mean is not the global mean, calculated by ignoring sex: 17.48. (Why?) The variance is estimated to be $\hat{\sigma}^2 = 60.03$, as compared with the value of $\hat{\sigma}^2 = 63.28$ obtained in Section 4.4.1 when differences in sex were ignored.

The AICs with and without sex difference are, respectively, 56.2 and 55.9, providing little indication of difference between the sexes.

The analysis of variance is given in Table 5.3. The F statistic, with 1 and 24 d.f., is not even significant at the 20% level. The corresponding Student t value for the difference between two means is 1.14 with 24 d.f., giving the same conclusion. (Notice that $1.14^2 = 1.30$.)

The normed profile likelihood for α_1 is plotted in Figure 5.7. The zero value is well within the range of plausible values of the parameter. Thus, we have no evidence of difference between the sexes in the time taken to do the homework, although the estimated mean time is longer for male than for female students. □

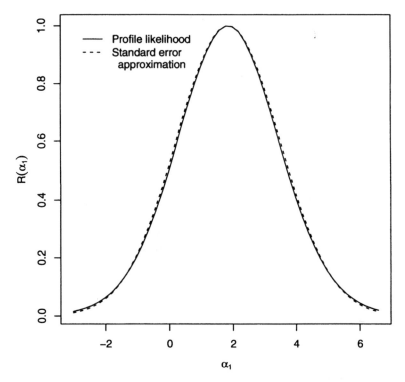

Fig. 5.7. Exact and approximate normed profile likelihoods for the difference of the males from the mean time of the two sexes to do the homework.

If the distribution of responses is skewed, it may be preferable to perform the analysis of variance on transformed values of the response (Section 4.6).

Example (continued)
Because we saw, in Section 4.4, that the overall distribution, ignoring sex, was asymmetric, we might also assume that this is true within each sex group and use the log normal distribution. This can be done by taking logarithms of the times (Section 4.6.1). The mean log times are estimated to be 2.853 and 2.606, respectively. These give $\hat{\mu} = 2.730$ and $\widehat{\alpha_1} = 0.124 = -\widehat{\alpha_2}$ in Equation (5.5), again using mean constraints.

Here the AICs with and without sex difference are, respectively, -17.7 and -17.8, again providing little indication of difference between the sexes.

The analysis of variance is given in Table 5.4. Again, the F statistic is not significant at the 20% level. The corresponding Student t value is 1.31 with 24 d.f. Thus, using logarithms, we also have no evidence of difference in the time taken to do the homework.

Table 5.4. Analysis of variance for the homework data in Table 5.2 using logarithms of the times.

	SS	MSS	d.f.	F
Total effect	5.66	0.226	25	
Sex effect	0.38	0.377	1	1.71
Residual	5.28	0.220	24	

For these data, transforming the response makes little difference to the inference conclusions obtained. Recall, however, that the interpretation of the parameter values is more difficult after transformation.   □

The assumption that the variance is constant can easily be checked in simple models like this by fitting a separate normal distribution to each subgroup. In this way, each subgroup will have a different mean and a different variance. Because the observations are independent, the AICs for the subgroups can be added and the result compared with those for the standard models described above.

Example (continued)
For the homework data, the estimated means and variances are, respectively, 2.853 and 0.165 for males and 2.606 and 0.265 for females (using logarithms of the times). Thus, the males take longer on average, but have a smaller variability. However, the sum of the two AICs is -17.1, indicating a slightly poorer model than either of those above. Although the estimated variances are quite different, there is no more indication that such a difference is necessary in the model than a difference in means.   □

5.3.2   TWO EXPLANATORY VARIABLES

As in Chapter 2, the generalisation of the linear model to two explanatory variables is direct:

$$\mu_{jk} = \mu + \alpha_j + \beta_k + \gamma_{jk}$$

with one of the usual sets of constraints. (Of course, this could also be specified as a multiple regression using sets of indicator variables.) Once again, we shall need to consider the two main effects, α_j and β_k, and the interaction between them, γ_{jk}. To simplify calculations here, I shall only look at balanced data where there are the same numbers of observations, $I = n_{jk}$, for each combination of the two explanatory variables, with total, $n_{\bullet\bullet} = \sum_j \sum_k n_{jk} = IJK$. Then, the maximum likelihood estimates are

Table 5.5. Four learning scores for Koerth pursuit rotor with 1 or 3 minutes rest and 3 or 5 practice sessions. (McNemar, 1954, p. 299, from Renshaw)

Rest	Number of practice sessions							
interval		3				5		
1	6	8	11	12	14	1	1	8
3	9	12	13	14	14	17	10	11

$$\hat{\mu} = \bar{y}_{\bullet\bullet\bullet}$$

$$\widehat{\alpha_j} = \bar{y}_{\bullet j\bullet} - \bar{y}_{\bullet\bullet\bullet}$$

$$\widehat{\beta_k} = \bar{y}_{\bullet\bullet k} - \bar{y}_{\bullet\bullet\bullet}$$

$$\widehat{\gamma_{jk}} = \bar{y}_{\bullet jk} - \bar{y}_{\bullet j\bullet} - \bar{y}_{\bullet\bullet k} + \bar{y}_{\bullet\bullet\bullet}$$

In this special balanced case, the maximum likelihood estimates do not change if we set one or more parameters to zero, for example, $\gamma_{jk} = 0$ for all j, k. Usually, this only makes sense if it is done in a *hierarchical* fashion; a main effect is not set to zero if the interaction is not zero.

For each of these possible models, the variance can be estimated using the appropriate estimates of the mean. For example, in the no-interaction model, we shall have

$$\widetilde{\mu_{jk}} = \hat{\mu} + \widehat{\alpha_j} + \widehat{\beta_k}$$

$$= \bar{y}_{\bullet j\bullet} + \bar{y}_{\bullet\bullet k} - \bar{y}_{\bullet\bullet\bullet}$$

$$\tilde{\sigma}^2 = \frac{1}{n_{\bullet\bullet}} \sum_i \sum_j \sum_k (y_{ijk} - \bar{y}_{\bullet j\bullet} - \bar{y}_{\bullet\bullet k} + \bar{y}_{\bullet\bullet\bullet})^2$$

The likelihood, AIC, and deviance can be calculated in the usual way, using Equations (5.2) and (4.7). However, as in the previous section, an analysis of variance table can also be set up:

	SS	MSS	d.f.	F
Total effect	$S_1 = \sum_{ijk}(y_{ijk} - \bar{y}_{\bullet\bullet\bullet})^2$	$M_1 = \frac{S_1}{n_{\bullet}-1}$	$n_{\bullet} - 1$	
Effect A	$S_2 = IK\sum_j(\bar{y}_{\bullet j\bullet} - \bar{y}_{\bullet\bullet\bullet})^2$	$M_2 = \frac{S_2}{J-1}$	$J - 1$	$\frac{M_2}{M_5}$
Effect B	$S_3 = IJ\sum_k(\bar{y}_{\bullet\bullet k} - \bar{y}_{\bullet\bullet\bullet})^2$	$M_3 = \frac{S_3}{K-1}$	$K - 1$	$\frac{M_3}{M_5}$
Interaction	$S_4 = S_1 - S_2 - S_3 - S_5$	$M_4 = \frac{S_4}{C}$	C	$\frac{M_4}{M_5}$
Residual	$S_5 = \sum_{ijk}(y_{ijk} - \bar{y}_{\bullet jk})^2$	$M_5 = \frac{S_5}{n_{\bullet}-JK}$	$n_{\bullet} - JK$	

where $C = (J-1)(K-1)$.

Example
Table 5.5 gives scores on a psychological test, called the Koerth pursuit rotor, used in the study of the acquisition of a motor skill. These are coded learning

Table 5.6. Analysis of variance for the learning scores of Table 5.5.

	SS	MSS	d.f.	F
Total effect	302.94	20.20	15	
Rest effect	95.06	95.06	1	6.09
Practice effect	5.06	5.06	1	0.32
Interaction	18.06	18.06	1	1.17
Residual	184.75	15.40	12	
Residual+Interaction	202.81	15.60	13	

scores, the sum of scores on the 29th and 30th trials in a study. Two different types of conditions were used. The rest interval was set at either one or three minutes. Those participants given three practice sessions had them on Monday, Wednesday, and Friday, whereas those with five had them on all week days.

Before we proceed, it is worth looking more closely at these data. We see that two people with one minute rest and five practice sessions only obtained a score of one. This should be investigated to verify that it is correct before performing any analysis. Although I am unable to do this, I shall nevertheless continue.

Here, $J = K = 2$ and there are $I = 4$ observations for each combination, for a total of $n_\bullet = IJK = 16$ scores. The four estimated means are $\bar{y}_{\bullet 11} = 9.25$, $\bar{y}_{\bullet 12} = 6.00$, $\bar{y}_{\bullet 21} = 12.00$, and $\bar{y}_{\bullet 22} = 13.00$, where j indexes the rest interval and k the number of practice sessions. Thus, the maximum likelihood estimates are

$$\hat{\mu} = 10.06$$
$$\widehat{\alpha_1} = -2.44$$
$$\widehat{\beta_1} = 0.56$$
$$\widehat{\gamma_{11}} = 1.06$$

The AIC for the full model is 24.6. When the interaction is set to zero, the AIC is 24.3. When $\alpha_1 = \gamma_{11} = 0$, that is, no rest effect, it is 26.4. On the other hand, when $\beta_1 = \gamma_{11} = 0$, no practice effect, the AIC is 23.5, indicating that this effect can be eliminated. Compared with this model, that with $\alpha_1 = \beta_1 = \gamma_{11} = 0$ has AIC 25.5, confirming that the rest effect cannot be removed.

The analysis of variance is given in Table 5.6. From Table A.3, we see that the P-value for the interaction effect is nonsignificant. When this occurs, it is customary to add together the sums of squares for interaction and residual, as is done in the last line of the table. (This is not strictly legitimate, as a basis for testing, because the decision to do so is based on the data.) Then, the mean sum of squares obtained from this is used in the denominator of the F statistics for the main effects. The two main effect F values in the table have been calculated in this way. We then have 1 and 13 d.f. for the test, showing that the rest effect is significant at the 5% level, whereas the practice effect is not.

The normed profile likelihood for the rest parameter α_1 is plotted in Figure 5.8. A value of zero is relatively implausible. Thus, less rest produces lower

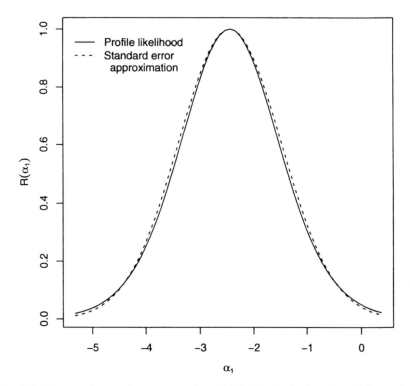

Fig. 5.8. Exact and approximate normed profile likelihoods for the effect of three minutes rest on learning scores, as compared with the mean of all scores.

scores (on average 2.44 points), with no evidence of a practice effect or an interaction between the two (given that the results in the table are correct).   □

5.3.3 MATCHED PAIRS

One important special design of studies involves pairs of observations that are matched for similarity in some way. The goal is to determine if there is any difference, on average, between pairs of responses, taking into account the variability among the pairs. One possibility is to apply a two-way analysis of variance. One explanatory variable is the pair and the other distinguishes the two responses of the pair. Thus, $I = 1$, $J = 2$, and K is the number of pairs. However, an equivalent simpler procedure is available. If we take the difference in the two responses of each pair, we can examine whether or not this differs from zero, on average. (Why does such a procedure only make sense in a location-scale family model?)

The AICs will compare a model having mean difference not equal to zero with one having zero mean. The appropriate test of significance will be a matched pairs

Student t test for a zero mean:

$$t_{n_\bullet - 1} = \frac{|\bar{y}_\bullet|}{\sqrt{(y_i - \bar{y}_\bullet)^2 / n_\bullet / (n_\bullet - 1)}}$$

where here y_i is the difference for pair i and $n_\bullet$ is the number of pairs. There is one less degree of freedom than the number of pairs. Notice that this is quite different from the model for the difference of two means when there is no matching in Section 5.3.1. Here, we look at the mean of the differences; with no matching, we look at the difference of means. The former allows for the variability among pairs; the latter cannot.

Example (continued)

Let us look again at the results of dieting in Table 1.7. The matching is between the two weights for each participant. As we saw in Section 5.2.1, the average change in weight is $\bar{y}_\bullet = -1.8$. The model with this constant difference is identical to that fitted above in Equation (5.4). Thus the AIC is 9.8, whereas that without a difference has 11.4. This latter model is the same as a linear regression model with $\beta_0 = 0$ and $\beta_1 = 1$. The AICs clearly indicate a difference after diet.

The Student t statistic is 2.5 with 9 d.f. The P-value is less than 0.025, showing some support for a difference in weight after diet.

The normed profile likelihood for the constant difference parameter β_0 is plotted in Figure 5.9. A value of zero is relatively implausible.

In Section 5.2.1, we saw that the model whereby weight after diet depends on the weight before fits somewhat less well than that with constant weight change for all participants that we have just re-examined.

The same conclusions would be reached by fitting a two-way analysis of variance without interaction, the two dimensions being the individuals and their two weights, and examining whether or not the effect of difference within pairs of weights is nonzero.   □

5.3.4   ANALYSIS OF COVARIANCE

Often, we shall wish to model responses that depend on both qualitative and quantitative explanatory variables, a combination of the linear regression and analysis of variance models. As we saw in Section 5.2.2, if we use indicator variables for the qualitative variable(s), we just have a complex multiple regression situation.

Example

The mean response for income may increase with the level of education, a quantitative variable, but the way in which this increase occurs may be different for the two sexes, a nominal variable. It might increase in the same way for both sexes, but be uniformly higher for one sex than the other. Or mean income might increase faster with increasing education for one sex than for the other, an interaction effect.   □

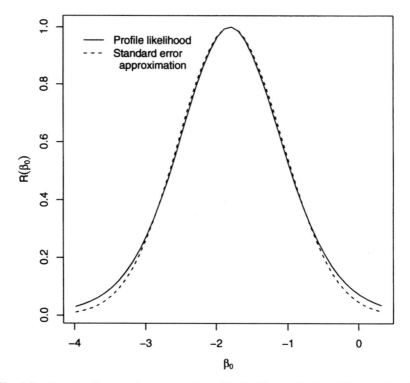

Fig. 5.9. Exact and approximate normed profile likelihoods for the difference in weight before and after diet.

Thus, it is possible to have a linear model with both nominal and quantitative explanatory variables. Such a model can be written in a number of ways (as well as by a multiple regression using indicator variables), including

$$\mu_{jk} = \beta_{0k} + \beta_{1k} x_{jk}$$
$$= \mu + \alpha_k + \beta_{1k} x_{jk}$$

where k indexes the nominal variable and j the quantitative variable. In this model, the regression line has a different slope β_{1k} for each category k of the nominal variable. This is illustrated in the upper left graph of Figure 5.10, with three categories for the nominal variable. Two slopes are positive and one negative. This is a model with interaction. The effect of the quantitative explanatory variable x_{jk} on the response variable is different in the different subpopulations, defined by the categories of the nominal explanatory variable.

One possibility is to fit this model as a separate regression for each subpopulation, using the methods of Section 5.2. Because observations are independent, the AICs, or deviances, can be added together for comparison with those of the

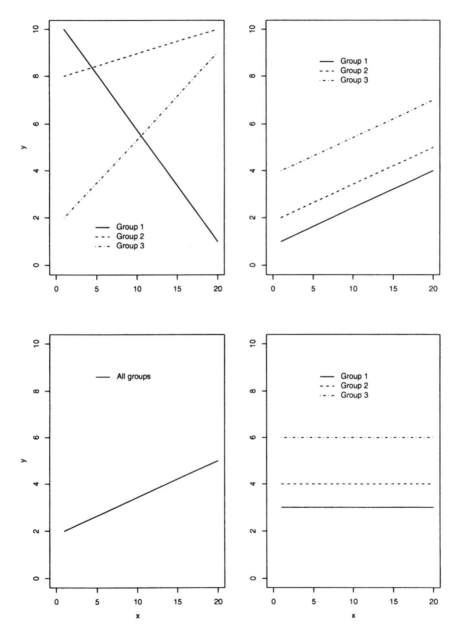

Fig. 5.10. Analysis of covariance with four different models.

simpler models to follow. However, this approach allows a different variance in each subgroup, something that does not occur if all regression lines are fitted simultaneously.

Example (continued)
We saw in Section 5.2.2 that the data on weights of babies in Table 5.1 vary with age in the same way for both sexes so that this interaction model was not necessary.   □

In a second model, all regression lines have the same slope so that they are parallel:

$$\mu_{jk} = \beta_{0k} + \beta_1 x_{jk}$$

as illustrated in the upper right graph of Figure 5.10. Here, both explanatory variables have effects, but they are additive. There is no interaction between them. The response distribution changes in the same way with x_{jk} within all groups, but at different levels. Here, we have

$$\widehat{\beta}_1 = \frac{\sum_j \sum_k (x_{jk} - \bar{x}_{\bullet k})(y_{jk} - \bar{y}_{\bullet k})}{\sum_j \sum_k (x_{jk} - \bar{x}_{\bullet k})^2}$$

$$\widehat{\beta}_{0k} = \bar{y}_{\bullet k} - \widehat{\beta}_1 \bar{x}_{\bullet k}$$

from which the variance, AIC, and deviance can be obtained in the usual way.

Example (continued)
This is the model for the weights of the babies using an indicator variable for sex and plotted in Figure 5.5.   □

If the response does not depend on the nominal variable, we just have one regression line,

$$\mu_{jk} = \beta_0 + \beta_1 x_{jk}$$

as in the lower left graph of Figure 5.10. The response distribution changes with x_{jk} in exactly the same way within all groups. This is the simple linear regression of Section 5.2, for all groups aggregated together.

Example (continued)
The simple regression model for the weights of the babies, ignoring sex, was plotted in Figure 5.2.   □

If the response depends on the nominal variable, but not on the quantitative regression variable, we shall have a different mean in each category, but this mean will not change with x_{jk},

$$\mu_{jk} = \beta_{0k}$$
$$= \mu + \alpha_k$$

as illustrated in the lower right graph of Figure 5.10. The response distribution has a different level in each group, but does not change with x_{jk} within a group. This is the analysis of variance model, with one explanatory variable, of Section 5.3.1.

Example (continued)
This is not an appropriate model for the weights of the babies because it assumes that they differ by sex, but do not change with age.   □

Finally, if the response does not depend on either variable, the mean is constant for all values of both,

$$\mu_{jk} = \beta_0$$

There is one common mean for all individuals, and we are back in the simplest situation, as in Section 4.4.1. A graph would just contain one horizontal line, the same for all groups.

5.4   Correlation

Let us suppose now that two continuous response variables have a *joint* distribution, as in Section 1.3.2.

Example
Consider the income of husband and wife pairs. Each income in the pair will have a distribution. As well, when one spouse has a relatively high income, there will be a good chance that the other will too. Thus, both incomes have a joint distribution such that they vary together in some way.   □

If the distributions are normal, this is called a *bivariate normal distribution*. Then, each response can have a different mean μ_j, describing its location, and a different variance σ_j^2, describing its variability. But there will also be a fifth parameter, the *correlation coefficient* ρ, describing the dependence between the two responses. The maximum likelihood estimate of this parameter is

$$\hat{\rho} = \frac{\Sigma_i (y_{1i} - \bar{y}_{1\bullet})(y_{2i} - \bar{y}_{2\bullet})}{\sqrt{\Sigma_i (y_{1i} - \bar{y}_{1\bullet})^2 \Sigma_i (y_{2i} - \bar{y}_{2\bullet})^2}}$$

This can take values in the interval, $[-1, 1]$.

Examples of bivariate normal observations are shown in the first three graphs of Figure 5.11, with independence $\rho = 0$, with positive correlation $\rho > 0$, and with negative correlation $\rho < 0$. With positive correlation, we see that, when one variable is greater than its mean, it is more probable that the other variable will

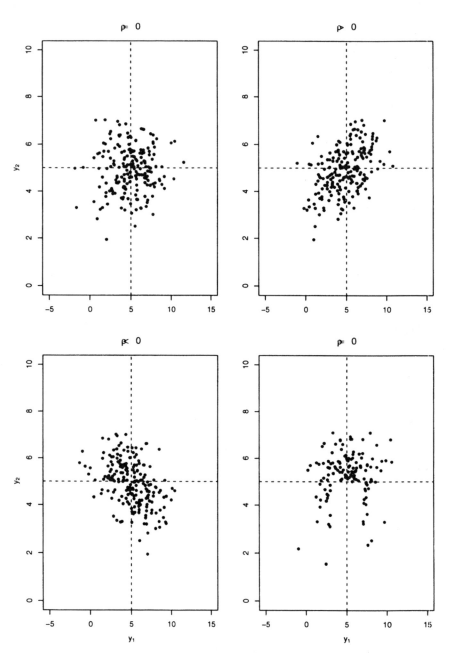

Fig. 5.11. Bivariate observations with means 5, variances 1 and 2, and various correlations. The lower right graph shows non-normal data.

also be greater than its mean. On the other hand, with negative correlation, when one variable is greater than its mean, it is more probable that the other variable will be smaller than its mean and vice versa. When there is zero correlation, no such prediction can be made.

The correlation parameter measures the dependence between the two variables. If $\rho = 0$, when they are normally distributed, then they are independent. As with any empirical mean or variance (Section 1.2.2), this value can be calculated for any two quantitative variables, even if they are far from being normally distributed. If they are independent, the parameter will be zero (or empirically close to it). But, conversely, *a value of zero cannot be interpreted to indicate independence unless the responses are normally distributed.* This is illustrated in the lower right graph of Figure 5.11. The correlation will be estimated to be just about zero, but there is an obvious nonlinear relationship between the two variables. Thus, great care must be taken in any interpretation of a correlation coefficient: it measures *linear* dependence with normal distributions.

5.5 Sample size calculation

As for logistic and log linear models in Section 3.6, we can calculate the sample size required to detect mean differences among normal distributions of responses in subgroups defined by explanatory variables. Here the task is complicated by the unknown variance in the model. To simplify matters, I shall assume that the variance is the same in the two models to be compared.

I shall only consider one special case of the simple one-way analysis of variance model, as in Section 5.3.1. We wish to distinguish a given sized difference of means in two equal-sized subgroups from zero. Let this difference be $\mu_1 - \mu_2 = 2\alpha_1$, as in Equation (5.5). The worst case situation will occur if a sample has half this difference. In this case, we have $\bar{y}_{\bullet 1} = \mu + \alpha_1/2$, $\bar{y}_{\bullet 2} = \mu - \alpha_1/2$, and $\bar{y}_{\bullet\bullet} = \mu$. Then,

$$R = \frac{(\sigma^2)^{-n_\bullet/2} e^{-\frac{1}{2\sigma^2} \Sigma_i \Sigma_j (y_{ij} - \bar{y}_{\bullet\bullet})^2}}{(\sigma^2)^{-n_\bullet/2} e^{-\frac{1}{2\sigma^2} \Sigma_i [(y_{i1} - \bar{y}_{\bullet 1})^2 + (y_{i2} - \bar{y}_{\bullet 2})^2]}}$$

$$= e^{-\frac{1}{2\sigma^2} [-n_\bullet \bar{y}_{\bullet\bullet}^2 + n_\bullet \bar{y}_{\bullet 1}^2/2 + n_\bullet \bar{y}_{\bullet 2}^2/2]}$$

$$-\log(R) = \frac{n_\bullet}{2\sigma^2} \left[\frac{\bar{y}_{\bullet 1}^2}{2} + \frac{\bar{y}_{\bullet 2}^2}{2} - \bar{y}_{\bullet\bullet}^2 \right]$$

$$= \frac{n_\bullet}{2\sigma^2} \left[\frac{(\mu + \alpha_1/2)^2}{2} + \frac{(\mu - \alpha_1/2)^2}{2} - \mu^2 \right]$$

$$= \frac{n_\bullet}{\sigma^2} \frac{\alpha_1^2}{8}$$

$$= \frac{n_\bullet}{\sigma^2} \frac{(\mu_1 - \mu_2)^2}{32}$$

$$n_\bullet = \frac{-32\log(R)\sigma^2}{(\mu_1 - \mu_2)^2}$$

Thus, we see that the sample size is directly proportional to the variance in the model and inversely proportional to the square of the difference between the means. Notice that this result does not depend on the overall mean (only the difference of means) whereas, for the corresponding logistic model, the sample size did.

Example

Suppose that we want to detect a difference in mean, $\mu_1 - \mu_2 = 6$, between two subgroups, say sexes, and we think that the variance may be about $\sigma^2 = 16$. As for the logistic model, we shall choose equal numbers of each sex for our sample. As in that example, if we use the AIC, the negative log normed likelihood will be unity. Then, the calculations are

$$n_\bullet = \frac{32 \times 1 \times 16}{6^2}$$
$$= 14.22$$

Thus, we would need a sample of at least 15 to detect this difference.
Suppose now that the variance was $\sigma^2 = 25$. We have

$$n_\bullet = \frac{32 \times 1 \times 25}{6^2}$$
$$= 22.22$$

and require a sample of 23.   □

Again, this method can be extended to more complex models, but with the difficulty of specifying values for a larger number of parameters.

5.6 Exercises

(1) The times in seconds for 30 children, classified by age, to push a hockey ball between a series of sticks during physical education were recorded (McPherson, 1990, p. 272) as shown in the following table:

Age	Time							
10	37	45	41	87	53	27	105	46
	27	35	38	54	19	36	30	
16	9	14	11	14	9	18	6	8
	30	8	10	12	16	23	14	

(a) Is there any indication of a difference in time between the two age groups?

(b) Do you think that the variability is the same in the two age groups?

(c) Why is a model for matched pairs not suitable for these data?

(2) The table below gives the percentage of eligible voters casting ballots in the 1964 Vancouver civic election and the mean income (dollars) in 1961 in 24 districts of the city (Erickson and Nosanchuk, 1977, p. 206, from Ewing).

District	Income	Turn-out
East side		
Cedar Cottage	3974	40
Collingwood	4186	38
Fraserview	4173	42
Grandview	3864	42
Kingsway	3865	38
Little Mountain	4383	43
Mt. Pleasant	3422	30
New Brighton	4003	39
Newport	4594	41
Riley Park	3865	38
Strathcona	2751	24
Sunset	4299	40
Woodland	3315	26
West side		
Arbutus	6267	55
Burrard	3589	27
Dunbar	5701	58
Fairview	3786	30
Kerrisdale	7066	59
Kitsilano North	3785	34
Kitsilano South	4558	41
Marpole	4640	41
Pt. Grey	5908	48
Shaughnessy	8477	52
West End	4233	33

The districts of Vancouver are distinct in the eyes of long-term residents of the city, and almost anyone who lives there knows them by name, although they have no administrative status.

(a) Study the relationship between the two variables, income and turn-out. Graphics will be useful.

(b) What reasons can you find to explain your results?

(c) Suggest a better model if the number of eligible voters in each district were available.

(d) The unit of observation is the district, each having a distinct geographical location. Is it reasonable to assume that such observations are independent?

    (e) Is this a sample, and is there a well-defined population?

    (f) Does it make sense to draw inferences about true parameter values in a model for data such as these?

(3) The table below presents the murder rates in 24 randomly chosen cities in the USA, classified by type and location (Blalock, 1972, p. 335).

Region	Industrial		Trade		Government	
			City type			
NE	4.3	5.9	5.1	3.6	3.1	3.8
	2.8	7.7	1.8	3.3	1.6	1.9
SE	12.3	9.1	6.2	4.1	6.2	11.4
	16.3	10.2	9.5	11.2	7.1	12.5

    (a) Study the effects of each of these two nominal variables separately on the observed murder rates.

    (b) Now construct a model to describe the simultaneous effects of the two explanatory variables.

    (c) Discuss the advantages of this second approach over the first.

    (d) Rates refer to the occurrence of events, here murders. If we had the numbers of such events in each city, what other information would we require in order to calculate the rates?

    (e) What distribution might be more appropriate than the normal, if such information were available?

(4) The data in Exercise (5.2) above are classified by the two main regions of Vancouver, the East and West sides. Construct models to compare

    (a) the mean income for the two regions;

    (b) the percentage turn-out for the regions.

(5) Construct a model to compare the relationship between voter turn-out and mean income for the two regions of Vancouver in Exercise (5.2) above.

(6) The following table gives the salaries (dollars) of board chairpersons of community organisations in the USA, classified by type of organisation and size of community (Blalock, 1972, p. 358).

Size of community	Religious	Social welfare	Civic
	Organisation type		
Large	13000	15000	20800
	11500	10600	18100
	17300	12300	18100
	19100	11400	22300
	16700	10800	16500
Small	15000	9300	14400
	12300	10400	10800
	13900	12900	9700
	14300	11000	12300
	11700	9100	13100

Five organisations of each type were randomly selected for both large and small communities. No further information is given about the definitions of the variables. Study the relationships between these two nominal variables and the salaries by fitting an appropriate model.

(7) The following table gives estimations of an index of the cost of living in five areas of Bengal, India, in 1945 by five investigators (Yule and Kendall, p. 529):

Investigator	A	B	C	D	E
			Area		
1	270	263	264	263	260
2	280	265	274	274	279
3	275	284	278	271	296
4	271	269	272	297	274
5	279	267	269	263	284

(a) Is there a difference in cost of living among the areas?

(b) Is there any evidence that the investigators differ in their evaluations of the areas?

(c) Why can you not determine if each investigator used the same criterion in all areas?

(8) In a study of 24 fifth-grade children at the School of Behavioural Sciences in Macquarie University, Australia, the time taken to solve four block design problems and the value for the embedded figures test (EFT), a measure of difficulty in abstracting logical structure of a problem from its context, were recorded. The children were classified by the type of problems presented first, those solved by row (group 1) or by formation strategy (group 2) as shown in the following table (Aitkin et al., 1989, p. 344):

Group	Time	EFT	Time	EFT	Time	EFT	Time	EFT
1	317	59	464	33	525	49	298	69
	491	65	196	26	268	29	372	62
	370	31	739	139	430	74	410	31
2	342	48	222	23	219	9	513	128
	295	44	285	49	408	87	543	43
	298	55	494	58	317	113	407	7

(a) Is there any relationship between the time taken on the block design problems and the results of the embedded figures test?

(b) Do the results for either of these measures differ with the order of presentation?

(c) Develop a complete model for these data and explain your conclusions.

(9) (a) Would any of the explanatory variables in any of the exercises above be susceptible to use as causal effects?

   (b) Do the above data tables provide such information?

   (c) How would you go about collecting information on causality in such
       contexts?

(10) Many of the data sets in the examples and exercises of this chapter may
     seem rather contrived. This illustrates the very real difficulty in finding
     data that might plausibly be described by models based on the normal dis-
     tribution. Which of these tables do you think contain purely invented data?

(11) (a) Calculate the sample size required to detect a difference in means of
          10 for a suitable value of the deviance.

      (b) Plot this as a function of the variance.

6
Dependent responses

As computers increase in power, more and more data are collected. When large quantities of data are available on the same individuals, they will often be inter-dependent in certain specific ways. Thus, more complex models must be used than those provided in the previous chapters. In an introductory text, I can only present a few of the basic concepts. Generally, one assumes that observations on the same individual may be dependent but that those on different individuals are independent.

Thus, dependent observations will arise primarily when several responses are recorded from the same individual. They may involve the same variable or several different response variables. Such responses may be obtained more or less simultaneously or they may evolve over time. When more than one individual is involved, if the same response variable is being measured several times on each, such observations are called *repeated measurements*.

In contrast, most of the models in previous chapters (except the log linear models of Section 2.3.2 and the bivariate normal distribution of Section 5.4) required responses that were independent. Suitable data were obtainable by observing independent individuals only once for each variable. In this way, the probabilities could be multiplied together to obtain the joint distribution of all observations. As well, all the models presented in the previous chapters were applicable to samples from clearly defined populations. In this chapter, I shall look at some processes, mentioned at the end of Section 1.3.1, for which a population of all possible responses is not necessarily well defined.

6.1 Repeated measurements

Any of the various types of responses covered in Chapter 4 may be recorded as repeated measurements.

- The simple presence or absence of an event, at each equally-spaced time point may be recorded as a series of, say, zeroes and ones, in a *point process* (Section 4.3.1). In more complex situations, the presence of one of a set of several possible events may be recorded at each instant.
- A count of the number of events may be recorded at each time point, as for the number of deaths from AIDS, or the number of unemployed, each month. This might be modelled as a Poisson process.

- Some continuous variable, such as the income of an individual, may be measured at each time point, in a *time series*.
- The duration times between successive unequally-spaced events, such as illness, strikes, or unemployment, may be of interest, called an *event history*.

Appropriate models should be adopted in each situation.

Example
In a study of election behaviour, the choice among political parties may be recorded each month for a sample of individuals. A study designed to follow individuals in this way is called a panel (Section 1.4.3). A model may be required to help understand how voters change among parties (such as those in Section 6.2.1). □

Multivariate models   When the data are dependent, the multivariate probability distribution of the responses on an individual must be modelled. This can be rather complex. Dependencies among responses following a multivariate normal distribution are given by their correlation coefficients (Section 5.4). Multivariate nominal responses, and counts in Poisson processes, can be modelled by log linear models, as we have seen in Section 2.3.2. Many other multivariate models are also available, but these are beyond the scope of this text.

Conditional models   Another approach is also possible. When the responses occur over time, we may expect that each succeeding response could depend on what type of responses preceded it. Thus, we can decompose the multivariate distribution using Equation (1.3):

$$f(y_1,y_2,y_3,\cdots) = f_1(y_1)f_2(y_2|y_1)f_3(y_3|y_1,y_2)\cdots \qquad (6.1)$$

Each factor in the product is a univariate (conditional) distribution. In this way, we are back in a situation with which we are familiar. We can use the models in Chapters 2 and 5 to study the dependence. Once the previous responses have been observed, they are just taken as explanatory variables in the part of the model involving the next response. We can simplify the model even further if we can assume the Markov property (Sections 4.3.2 and 4.5.2): each response only depends on that occurring immediately before it. Then, we have a *Markov process*,

$$f(y_1,y_2,y_3,\cdots) = f_1(y_1)f_2(y_2|y_1)f_2(y_3|y_2)\cdots \qquad (6.2)$$

Notice that, here, in contrast to Equation (6.1), we generally use the same density function $f_2(y_t|y_{t-1})$ for all responses except the first. Often, only this conditional distribution is modelled, because it is of prime interest, and the distribution of the first response $f_1(y_1)$ is ignored.

Even in more complex cases, it is usually only necessary to condition on responses two or three points back in time.

Cluster models   Another characteristic of repeated measurement data, whether measured over time or not, is that the different individuals observed may inherently react differently in the same situation. Explanatory variables will account for part of these differences, but rarely all. For example, for a quantitative response, some individuals will consistently give higher values than others, under all conditions. This is variously known as *proneness* (Section 4.3.1) or *frailty* if the causes are internal and *liability* if they are external (for example, environmental). Often, these cannot be distinguished and are globally called *susceptibility*.

In this chapter, I shall briefly introduce a few of these models for various types of dependencies, using only the techniques already discussed in previous chapters.

6.2 Time series

Any set of observations observed over time can be called a *time series*. The observations may simply be indicators of whether or not an event has occurred or they may be measurements of some quantitative response variable. Usually, observations are made at equally-spaced time intervals.

6.2.1 MARKOV CHAINS

A point process (Section 4.3.1) can be recorded as a series of 1/0 values indicating whether or not an event occurred at each time point. We may be interested in knowing if the occurrence of an event is more probable when there was already an event at the previous time point. In more complex situations, we may have an indicator of which of several types of events occurs at each time point. In a simple model, each response is assumed to depend only on the immediately preceding one. Such a Markov process for nominal variables observed at equally-spaced time intervals is called a *Markov chain*.

This process can be studied by using the response at that previous point as an explanatory variable. The easiest way to construct such a model is by using a *lagged* variable. The series of responses is displaced by one time point to form a new series, the explanatory variable.

Example
Consider the point process

$$11001011100001 \cdots \qquad \text{response variable}$$
$$11001011100001 \cdots \qquad \text{explanatory variable}$$

Below each response, we have what happened just before it. Notice that, without further assumptions, we shall not be able to use the first response, because we do not know what happened before it. In other words, in Equation (6.2), we would need to define $f_1(y_1)$ as well as $f_2(y_t|y_{t-1})$ if we want to use the first value.   □

Table 6.1. Days with accidents (1) for one shift in a mine (read across rows). (Lindsey, 1992, p. 21, from Maguire *et al.* 1952)

1	0	0	1	0	0	0	0	0	0	0	0	0	0	0	0	0	0	0
0	0	0	0	0	0	0	1	0	1	0	0	1	0	1	1	1	1	1
0	0	1	0	1	1	0	1	0	1	0	0	1	0	0	1	0	0	0
1	0	1	0	0	1	0	0	0	0	1	0	1	0	0	0	1	0	0
0	0	1	0	0	0	1	0	1	1	1	1	1	0	0	0	0	0	0
0	1	0	1	0	1	0	1	0	1	0	1	0	0	0	0	0	0	0
1	0	0	0	0	0	0	0	1	0	1	1	1	0	0	0	0	0	0
0	1	1	0	0	0	0	0	0	0	1	0	0	1	0	1	1	1	0
0	1	0	1	0	0	0	1	0	0	1	0	0	0	0	0	0	0	0
0	0	0	1	0	0	0	0	0	0	0	0	0	1	0	0	0	1	0
0	1	0	1	0	1	0	0	0	0	0	0	0	0	0	0	0	0	0
1	1	0	0	0	1	0	0	0	0	0	0	0	1					

Based on the response variable and its lagged values, we can create a 2×2 table with the four possible combinations, event or no event followed by event or no event. We can represent these relationships by

$$\begin{pmatrix} \pi_{0|0} & \pi_{1|0} \\ \pi_{0|1} & \pi_{1|1} \end{pmatrix}$$

where, for example, $\pi_{0|1}$ is the conditional probability of no event following an event. This is called a *transition matrix*, because it describes the changes between the time points. For a Markov chain, such a matrix completely defines $f_2(y_t|y_{t-1})$ in Equation (6.2).

I shall assume that the transition matrix does not change over time, a property called *stationarity*. In other words, $f_2(y_t|y_{t-1})$ is the same distribution at all time points.

The observations for such a transition matrix are just a special case of the tables to which we applied a logistic model in Chapter 2, and in particular, in Section 2.2.2. Thus, if we fit the model of Equation (2.4) and found that α_j could be zero, responses would not depend on what happened at the previous time point.

Example
Table 6.1 gives a binary series indicating whether or not at least one accident occurred in one section of a mine on one shift on each of 223 successive days. The year from which the data come is not stated. A useful plot of such data is the cumulative number of events against time, as shown in Figure 6.1. This gives an indication of the rate or intensity with which events are occurring. Here we see that the line is fairly straight, indicating a reasonably constant intensity, except for a period without accidents near the beginning.

If we create the lagged variable and count the number of occurrences of each combination of pairs of possible responses, we obtain the contingency table with

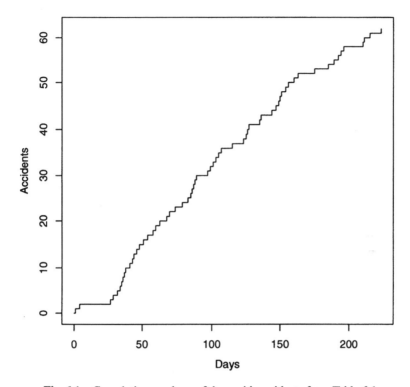

Fig. 6.1. Cumulative numbers of days with accidents from Table 6.1.

Table 6.2. Two-way contingency table created from Table 6.1.

	Response	
Lagged	0	1
0	115	46
1	46	15

the first day eliminated, given in Table 6.2. From this, the estimated transition
matrix of conditional probabilities is

$$\begin{pmatrix} \widehat{\pi_{0|0}} & \widehat{\pi_{1|0}} \\ \widehat{\pi_{0|1}} & \widehat{\pi_{1|1}} \end{pmatrix} = \begin{pmatrix} 0.714 & 0.286 \\ 0.754 & 0.246 \end{pmatrix}$$

There are many more sequences of two days without accident (00) than of two
with (11), so that the probability of (01) is estimated as 0.286 whereas that of
(10) is 0.754. However, the conditional probability of no accident is about the
same whether an accident occurred the previous day (0.754) or not (0.714). This
is confirmed by the AIC of 1.18 with 1 d.f. for the model of independence as

Table 6.3. A three-way contingency table created from Table 6.1.

		Lag 2		
	0		1	
		Response		
Lag 1	0	1	0	1
0	87	28	28	18
1	38	7	7	8

compared with 2 for the Markov dependence. The parameter estimates of the logistic model are $\hat{\mu} = 1.018$ and $\widehat{\alpha_1} = -0.102$.   □

The response can also be lagged two or more time periods to see if the response depends on what happened further back in time. Each new explanatory variable, created in this way, adds a further dimension to the table, and more parameters to the model. We might, thus, look at the response lagged two time periods before, in a second-order Markov chain model. This will yield a three-way contingency table.

Example (continued)
The contingency table for a second-order Markov chain model for the mine accidents, with the first two days eliminated, is shown in Table 6.3. The one period transition matrices are

$$\begin{pmatrix} \widehat{\pi_{0|00}} & \widehat{\pi_{1|00}} \\ \widehat{\pi_{0|01}} & \widehat{\pi_{1|01}} \end{pmatrix} = \begin{pmatrix} 0.757 & 0.243 \\ 0.844 & 0.154 \end{pmatrix}$$

between two days, when the previous day is without accident, and

$$\begin{pmatrix} \widehat{\pi_{0|10}} & \widehat{\pi_{1|10}} \\ \widehat{\pi_{0|11}} & \widehat{\pi_{1|11}} \end{pmatrix} = \begin{pmatrix} 0.609 & 0.391 \\ 0.467 & 0.533 \end{pmatrix}$$

when there was an accident. There is a considerable difference between these two matrices. Thus, the probability of an accident is higher when there was one two days before, and especially if there was on both previous days.

When we fit the two-variable logistic model, we obtain the parameter estimates, $\hat{\mu} = 0.783$, $\widehat{\alpha_1} = 0.004$ (effect of the previous day), $\hat{\beta}_1 = 0.629$ (effect two days back), and $\widehat{\gamma_{11}} = -0.283$ (interaction between the two previous days). The AIC is 7.6 if we only allow dependence on one day back, that is, have $\alpha_1 = \gamma_{11} = 0$, with 2 d.f. as compared with 4 for the full model; it is 3.2 on 2 d.f. with dependence only two days back, that is, $\beta_1 = \gamma_{11} = 0$. This confirms that there appears to be dependence if we look back two days in time.   □

As in Chapter 2, the same types of models can be used when the response variable is nominal, with more than two possible values at each time point. The transition matrix will still be square, but will be larger, with dimensions equal to the number of possible categories of response.

Example (continued)
Table 1.8 gave the place of residence of a sample of people in Britain in 1966 and 1971. Here, we have four possible events at each of the two time points. Thus, for example, 12 people in the sample who lived in Central Clydesdale in 1966 had moved to Lancashire or Yorkshire by 1971. The estimated transition matrix is

$$
\begin{pmatrix}
\widehat{\pi_{1|1}} & \widehat{\pi_{2|1}} & \widehat{\pi_{3|1}} & \widehat{\pi_{4|1}} \\
\widehat{\pi_{1|2}} & \widehat{\pi_{2|2}} & \widehat{\pi_{3|2}} & \widehat{\pi_{4|2}} \\
\widehat{\pi_{1|3}} & \widehat{\pi_{2|3}} & \widehat{\pi_{3|3}} & \widehat{\pi_{4|3}} \\
\widehat{\pi_{1|4}} & \widehat{\pi_{2|4}} & \widehat{\pi_{3|4}} & \widehat{\pi_{4|4}}
\end{pmatrix}
=
\begin{pmatrix}
0.738 & 0.075 & 0.044 & 0.144 \\
0.006 & 0.902 & 0.036 & 0.055 \\
0.003 & 0.025 & 0.933 & 0.039 \\
0.002 & 0.014 & 0.011 & 0.973
\end{pmatrix}
$$

giving the conditional probability for that particular move to be 0.075. As might be expected, the conditional probabilities of residing in the same place at the two points in time five years apart, given on the main diagonal of the matrix, are much higher than those of being elsewhere. Place of residence in 1971 depends very strongly on that in 1966 (usually being the same!). The AIC for the independence model is 9946 with 9 d.f., as compared with 14 for the Markov dependence, confirming this. □

   In the first example, one 'individual' (the shift in a mine) was observed a large number of times, whereas, in the second example, a large number of individuals was observed only two times each. In the first case, I have assumed that the transition matrix stays the same over time (stationarity) whereas, in the second, I assumed that it is the same for all individuals.

6.2.2 AUTOREGRESSION

Models similar to those for Markov chains can be constructed when the response is quantitative. In order to be able to use the models of Chapter 5, I shall assume that it is a normally-distributed continuous variable being observed over time. Once again, a new explanatory variable is constructed by lagging the response variable. Then, a linear regression can be applied, in the simplest cases, without even an intercept:

$$\mu_t = \rho y_{t-1}$$

where β_1 in Equation (5.1) is here called ρ. This Markov process is called an autoregression for obvious reasons. Here, the slope parameter ρ is also called the *autocorrelation*. If $|\rho| < 1$, the time series is stationary; it will be reasonably stable over time. An important special case of this model occurs when $\rho = 1$, a *random walk*, which is not stationary. If an intercept, β_0, is included, it gives the *drift* of the process.
   When there is no drift, the equation for the maximum likelihood estimate of the slope, the autocorrelation, simplifies to

$$\hat{\rho} = \frac{\sum_t y_{t-1} y_t}{\sum_t y_{t-1}^2}$$

Table 6.4. Annual percentage increase in average wages of white collar workers in Britain, 1962–1979. (Lindsey, 1992, p. 122, from Nichols, 1983)

1961–62	2.8	1970–71	6.2
1962–63	2.7	1971–72	6.3
1963–64	2.7	1972–73	5.5
1964–65	2.2	1973–74	6.2
1965–66	2.9	1974–75	9.1
1966–67	4.5	1975–76	7.6
1967–68	5.1	1976–77	6.9
1968–69	5.5	1977–78	7.5
1969–70	6.2	1978–79	7.2

This resembles the equations for estimating both a regression coefficient (Section 5.2.1) and a correlation coefficient (Section 5.4).

This model implicitly involves the construction of a multivariate normal distribution, the $f(y_1, y_2, y_3, \cdots)$ on the left of Equation (6.2), produced from the product of conditional, univariate normal distributions. Then, the dependence, as measured by the correlation among consecutive responses, is given by the autocorrelation ρ. The correlation between any pair of responses, not just adjacent ones, can also be obtained. It is given by $\rho^{\Delta t}$, where Δt is the distance in time between them. Thus, if $|\rho| < 1$, the correlation among pairs of responses decreases with increasing distance between them.

Example
Consider the series of percentage increases in salaries in Britain, given in Table 6.4, as reported in the *National Survey of Professional, Administrative, Technical, and Clerical Pay* for March, 1979. These are for jobs equivalent to low level government positions, rated GS1 to GS4, and are March to March changes.

For quantitative variables, it is always useful first to plot (Section 1.2.3) the data over time. This is done in Figure 6.2, for the data from Table 6.4. We see how the wage increases are generally growing over these years.

The autoregression model, with drift, has estimates, $\widehat{\beta_0} = 1.038$, $\widehat{\rho} = 0.853$, and $\hat{\sigma}^2 = 0.852$. The model without drift is very close to a random walk: $\widehat{\rho} = 1.025$, and $\hat{\sigma}^2 = 0.984$. From Equation (4.7), the deviance comparing these two models is $17\log(0.984/0.852) = 2.45$ with 1 d.f. The AICs are, respectively, 0.6 and 0.9. In contrast, the independence model, with $\rho = 0$, has $\hat{\sigma}^2 = 3.693$, so that the deviance, compared with this latter model, is $17\log(3.693/0.984) = 22.49$ with 1 d.f. and the AIC 12.1. The random walk model is about as plausible as the autoregression with drift, but independence is unlikely. □

With such time series data, another possibility is to fit a simple linear regression of the response variable on the time. This will show whether or not there is a *trend* in the data.

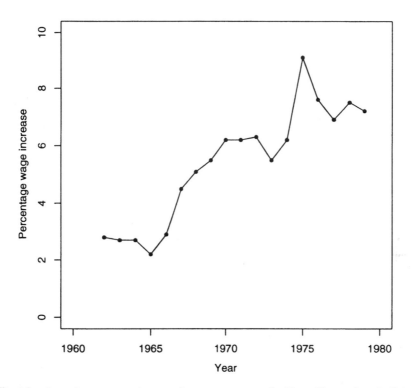

Fig. 6.2. Annual percentage increase in average wages of white collar workers in Britain, 1962–1979, from Table 6.4.

Example (continued)
For the salary data, let us number the years with consecutive integers and use this as an explanatory variable. (Here, I do not need to leave out the first year, but shall do so, for comparability with the previous results.) The estimated parameters are $\widehat{\beta}_0 = 2.040$, $\widehat{\beta}_1 = 0.351$, and $\widehat{\sigma}^2 = 0.741$, where here β_1 is the dependence of the response on time. The percentage increase in wages is estimated to be increasing by about 0.35 per year. This model has a smaller variance than any of the others, indicating that it fits better. The AIC is -0.55 and the deviance, compared with independence, is $17 \log(3.693/0.741) = 27.31$ with 1 d.f.

As we have already seen from Figure 6.2, this time series is not stationary, but is increasing with time. □

6.3 Clustering

When several observations are made on each of a number of individuals, those on each given individual will not generally be independent. We have now seen some ways in which they can be related over time. However, even if the observations were made a very long time apart, dependence could still be present. This is the proneness or frailty of certain individuals. Two randomly chosen responses from one individual will often be more similar than two randomly chosen from two different individuals.

Clustering is built into many study designs (Section 1.4.2), for example, when the unit of random sampling is a group, such as a class-room, family, or village, but several members of the group are questioned. Another important application of clustering is to meta-analysis. This involves the secondary analysis of a series of similar studies in order to combine the information from them. Responses within a given study may be expected to be more similar than those across different studies, so that each study is the 'individual' forming a cluster.

One way to model clustering is by letting the parameter of the probability density function vary across the population, a regression model. However, often appropriate explanatory variables are not available to account for all this variation. Then, an alternative is to allow the parameter itself to vary randomly, instead of systematically, following some probability distribution. This yields a *random effects* model.

Examples

In Sections 4.3.3 and 4.3.4, we saw some ways in which count data exhibit clustering through overdispersion. This is characteristic of such data, as opposed to frequency data, because the former involve a count of events on the same individual. Thus, if counts might follow a Poisson distribution, but all the individuals in the population do not have the same Poisson mean, this mean parameter might be allowed to vary according to a gamma distribution (Section 4.5.4). The resulting model is a negative binomial distribution, which we fitted to clustered data in Section 4.3.4.   □

Random effects models can also be developed for a normal distribution. Let us consider an analysis of variance situation, where we have several response observations for each individual i and one nominal explanatory variable, indexed by j, distinguishing among these responses. A simple model, without interaction, would be

$$\mu_{ij} = \mu + \alpha_i + \beta_j$$

This looks like a two-way analysis of variance (Section 5.3.2), where each response follows an $N(\mu_{ij}, \sigma^2)$. However, such a model would involve a separate parameter (α_i) for each individual i, possibly a considerable number.

Instead, let us assume that the parameter α_i, describing the differences among the individuals (using the mean constraint so that their mean is zero), varies randomly in the population following another normal distribution, $N(0, \sigma_c^2)$. In this way, we can replace all the values of α_i by one new parameter, σ_c^2. Remember,

Table 6.5. Judgement of three different levels of albedos by four observers at illumination level 2. (McNemar, 1954, p. 321, from Taubman)

	Level of albedos		
Observer	0.07	0.14	0.26
1	14	24	65
2	27	36	47
3	18	24	62
4	24	59	84

however, that this requires the strong assumption that the values of α_i follow a normal distribution with a constant variance.

The resulting model will be a multivariate normal distribution for the set of responses from each individual, where the correlation between any pair of that set of responses is constant, given by

$$\rho = \frac{\sigma_c^2}{\sigma_c^2 + \sigma^2}$$

This is called the *intraclass correlation*, because it describes the constant dependence among all responses within a class, here those of the same individual.

To set up the model, we can create a special analysis of variance table:

	SS	MSS	d.f.	F
Individual	$S_1 = J\sum_i(\bar{y}_{i\bullet} - \bar{y}_{\bullet\bullet})^2$	$M_1 = \frac{S_1}{I-1}$	$I-1$	
Effect A	$S_2 = I\sum_j(\bar{y}_{\bullet j} - \bar{y}_{\bullet\bullet})^2$	$M_2 = \frac{S_2}{J-1}$	$J-1$	$\frac{M_2}{M_3}$
Residual	$S_3 = \sum_{ij}(y_{ij} - \bar{y}_{\bullet j} - \bar{y}_{i\bullet} + \bar{y}_{\bullet\bullet})^2$	$M_3 = \frac{S_3}{C}$	C	

where $C = (I-1)(J-1) = IJ - I - J + 1$. From this, we have $\hat{\sigma}^2 = M_3$ and $\hat{\sigma}^2 + I\hat{\sigma}_c^2 = M_1$, so that $\hat{\sigma}_c^2$ and $\hat{\rho}$ can be calculated. (These are conditional maximum likelihood estimates, as mentioned in Section 4.4.1.)

Example
Table 6.5 gives the judgements about three different levels of albedos, that is, whiteness, by four observers. The scores in the table are the judged level of whiteness. Illumination was set at 2 in unspecified units. Because each observer gave a response for each albedos level, the values are not independent.

The analysis of variance table is

	SS	MSS	d.f.	F
Individual	950.0	316.7	3	
Albedos level	3954.2	1977.1	2	17.9
Residual	662.5	110.4	6	

This yields the variance estimates, $\hat{\sigma}^2 = 110.4$ and $\hat{\sigma}_c^2 = (316.7 - 110.4)/4 = 51.6$, so that the estimate of the intraclass correlation is $\hat{\rho} = 51.6/(51.6 + 110.4) = 0.32$. We see that there is considerable variation in the general levels of responses of the individuals, with the fourth observer giving particularly high answers. There also appear to be systematic differences for all observers among levels of albedos: the estimates are $\widehat{\beta_1} = -19.58$, $\widehat{\beta_2} = -4.58$ and $\widehat{\beta_3} = 24.17$ for levels 0.07, 0.14, and 0.26 respectively, with $\hat{\mu} = 40.33$. The F statistic for them being zero is significant at the 5% level.   □

One may ask what happens if repeated responses on individuals are more heterogeneous than across individuals. From the formula $\hat{\sigma}_c^2 = (M_1 - M_3)/I$, so that the intraclass correlation could be negative. This could happen if there is some sort of repulsion among events on an individual so that they are forced to be quite different (analogous to underdispersion for count data described in Section 4.3.1).

In more complex models, more explanatory variables will be available. As well, in longitudinal data, both time dependence and clustering will usually be present. However, these models require sophisticated computer software for their analysis.

6.4   Life tables

In Section 4.5, we studied probability distributions for duration times. One important tool, in this context, is the intensity function (Section 4.5.1). This describes how the risk, sometimes called the hazard, of an event changes over time. In this section, I shall show how to construct some simple models for the intensity function, and the corresponding survivor curve, of duration data, even when censoring is present.

6.4.1   ONE POSSIBLE EVENT

The basic idea in the construction of a life table is to follow individuals over time, recording, in each of a series of periods, the number who have some event, such as dying, finding a job, and so on. In the simplest case, each individual who has an event then disappears from the process. In such a situation, the event is said to be *absorbing*. Thus, in each successive time period, we have several quantities available:

(1) the number of individuals still susceptible to have the event at the beginning of the period (thus not including those who have already had the event or who have previously disappeared due to censoring), called the *risk set*;

(2) the number having the event in the period, among those at risk; and

(3) the number at risk disappearing (that is, censored) without an event in the period.

In each period or interval of time, indexed by t, an individual under observation (at risk) either has an event or not. We can model this process by a binomial distribution with the conditional probability, π_t^I, of having the event in the present interval, given that no event has occurred in any previous period. For this, we can use the relationship in Equation (6.1), so that the probabilities can be multiplied together. This series of probabilities forms the intensity function. These probabilities can be estimated in the usual way for binomial probabilities within each time period.

For simplicity, I shall assume that events occur during the period, whereas censoring occurs at the end of the period. Then, the conditional probabilities can be estimated by the number of events divided by total number under observation in each period. The problem of censoring is handled by this simple device in all that follows.

Thus, if there are n_t individuals under observation (at risk) in period t, of whom d_t have an event, then

$$\hat{\pi}_t^I = \frac{d_t}{n_t}$$

From these values, other useful probabilities can also be calculated. The *cumulative survivor probability*, π_t^S, of not having an event during the first t periods (Section 4.5.1) is given by multiplying together the corresponding conditional probabilities of not having events in each of the periods:

$$\pi_t^S = \prod_{i=1}^{t} (1 - \pi_i^I)$$

A table of estimates of these cumulative probabilities is called a life table and the method is called *Kaplan–Meier estimation*. The plot of these estimates is called the survivor curve.

Example
Table 6.6 gives data on the survival of two groups of women with different stages of cancer of the cervix. This is a cohort study (Section 1.4.3) whereby the women were examined only once a year. Censoring occurred if they ceased to attend the clinic where the study was conducted. Notice how the reduction in the number of women under observation (at risk) between any two years is equal to the sum of the number of events (deaths) and the number censored.

For the moment, I shall only look at those with Stage 1 cervical cancer. The estimates of the intensity and the Kaplan–Meier (binomial) cumulative probabilities are given in Table 6.7. As time proceeds, fewer individuals are under observation, so that the later estimates of probabilities will be less precise (the likelihood function will be wider). Because of this, the estimates of the intensity function vary rather irregularly. For this reason, the estimates of the cumulative survivor probabilities, given in the third column of the table, may be more useful. The survivor curve is plotted in Figure 6.3. Because 24 women were known still to be alive after the ten years, the survivor curve does not descend to zero. (Some of the censored women might also be alive.)   □

Table 6.6. Survival over a ten-year period of women with two stages of cancer of the cervix. (Clayton and Hills, 1993, p. 32)

	Stage 1			Stage 2		
Years	Number	Deaths	Censored	Number	Deaths	Censored
1	110	5	5	234	24	3
2	100	7	7	207	27	11
3	86	7	7	169	31	9
4	72	3	8	129	17	7
5	61	0	7	105	7	13
6	54	2	10	85	6	6
7	42	3	6	73	5	6
8	33	0	5	62	3	10
9	28	0	4	49	2	13
10	24	1	8	34	4	6

Table 6.7. Estimated intensities and cumulative survivor probabilities for the women suffering from Stage I cervical cancer in Table 6.6.

Years	Intensity	Binomial survivor	Poisson survivor
1	0.045	0.955	0.956
2	0.070	0.888	0.891
3	0.081	0.816	0.821
4	0.042	0.782	0.788
5	0.000	0.782	0.788
6	0.037	0.753	0.759
7	0.071	0.699	0.707
8	0.000	0.699	0.707
9	0.000	0.699	0.707
10	0.042	0.670	0.678

If we look at the pair of series, d_t and $n_t - d_t$, these data, from which a life table is constructed, form a two-way contingency table such as we studied in Chapter 2. The logistic model for independence in such a table

$$\log\left(\frac{\pi_t^I}{1 - \pi_t^I}\right) = \beta_0$$

assumes a constant intensity over all periods, that is, a Poisson process, corresponding to an exponential distribution of times between events (Sections 4.3.1 and 4.5.2). A second simple model allows the intensity to vary over time in a logistic regression,

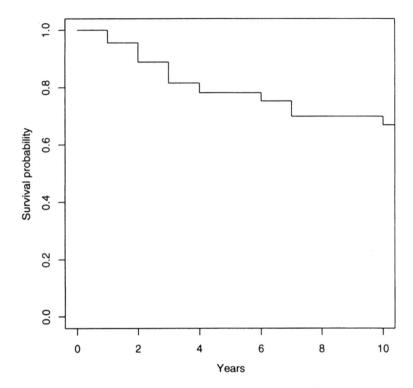

Fig. 6.3. Survivor curve for the cancer data of Table 6.6.

$$\log\left(\frac{\pi_t^I}{1 - \pi_t^I}\right) = \beta_0 + \beta_1 x_t$$

where x_t might be, for example, time or the logarithm of time.

Example (continued)
The data in Table 6.6 are reconstructed as a contingency table in Table 6.8. For the moment, I shall only consider the top half, for Stage 1 women. The AIC for the independence model, where the intensity does not depend on the number of years, is 8.7 with 9 d.f., as compared with 10 for the saturated model with complete dependence. This gives little evidence that the intensity is not constant over time.

The saturated model allows the intensity to change irregularly at each time point, and corresponds to the series of estimates of the intensity in Table 6.7. This can be simplified by fitting a logistic regression on time, which gives an estimate of the slope, $\widehat{\beta}_1 = -0.129$. (The approximate method of parameter estimation in Section 2.2.5 does not work when the table contains zeroes.) The AIC is 8.3 as

Table 6.8. Reconstructed contingency table for the women of Table 6.6.

Years	1	2	3	4	5	6	7	8	9	10
				Stage 1						
Alive	105	93	79	69	61	52	39	33	28	23
Dead	5	7	7	3	0	2	3	0	0	1
				Stage 2						
Alive	210	180	138	112	98	79	68	59	47	30
Dead	24	27	31	17	7	6	5	3	2	4

compared with 8.7 for a constant intensity. Thus, there is only a slight indication that the intensity is decreasing with time.   □

If explanatory variables are present, so that the table has more dimensions, various logistic models can still be fitted. If time is used as a categorical variable and it has no interactions with the other explanatory variables, we have a *Cox proportional hazards model*.

Example (continued)
Let us now look at both stages of cancer in Table 6.6, as reconstructed in Table 6.8. If we fit a logistic model to the whole table, we find that the interaction parameters, between time and stage, are not very large, so that the Cox model may be suitable. The estimate of the parameter for differences in stage is 0.935, indicating a higher intensity for Stage 2 than Stage 1. The deviances for independence of intensity from time and from stage are, respectively, 36.2 with 18 d.f. and 28.9 with 10 d.f. However, the AIC for the complete model is 20, whereas that without time is 20.1 and that without stage 24.5, so that intensities may be constant over time but are different between stages.
One way to check this is to replace the categorical variable for time by a quantitative variable. Thus, if we use a regression model, as above, but now introduce stage as well (as I did for the weight and sex of babies in Section 5.2.2), the AIC is 16.7. As in that example, an interaction is not necessary (AIC 17.6). Thus, the intensity (risk of cancer) is decreasing over time ($\widehat{\beta}_1 = -0.103$), but is consistently higher in Stage 2 ($\widehat{\beta}_2 = 0.923$).
For information about the precision of the two regression parameters, the normed profile likelihoods for β_1 and β_2 are plotted in Figure 6.4. Both indicate fairly precise estimates of the parameters.   □

6.4.2 REPEATED EVENTS
If more than one event can occur to each individual, as in event histories, the binomial distribution should be replaced by the Poisson. Then, from Equation (4.2), the probability of no event in a period becomes $e^{-\mu t}$ and that of one event, $\mu_t e^{-\mu t}$, where, as we saw in Section 4.3.1, μ_t is now the intensity of the event.

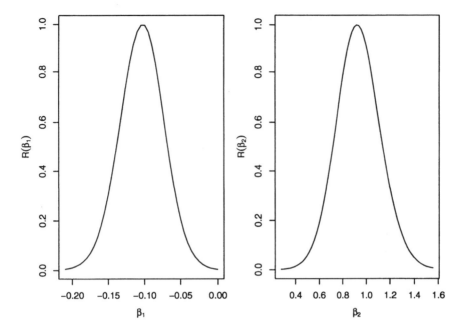

Fig. 6.4. Normed profile likelihoods for the slope parameters for dependence of risk of dying of cervical cancer on time and stage.

Whether individuals have zero, one, or more events in a period, the likelihood function is

$$L(\mu_t) = e^{-n_t \mu_t} \mu_t^{d_t}$$

where d_t is now the total number of events in period t, instead of the number of individuals with events in that period. The maximum likelihood estimate for the intensity in a period is

$$\hat{\mu}_t^I = \frac{d_t}{n_t}$$

Notice that this is the same as for $\hat{\pi}_t^I$ above, but now each individual may have more than one event in a period so that the estimate of the intensity does not necessarily lie between zero and one. Here, the cumulative probability of survival is again obtained by multiplying together the conditional probabilities of not having an event in the series of periods,

$$\pi_t^S = \prod_{i=1}^{t} e^{-\mu_i}$$
$$= e^{-\Sigma \mu_i}$$

where $\sum \mu_i$ is called the *cumulative intensity*. This procedure yields the *Aalen–Nelson estimates* of the life table or survivor curve.

If the periods are short enough so that the probability of an event in any period is small, these two different estimates of π_t^S will be similar. Thus, when the periods are short, the Aalen–Nelson estimate can be used as an approximation to the Kaplan–Meier estimate for survival studies, where only one event per individual is possible.

Example (continued)
For the data in Table 6.6, the Aalen–Nelson (Poisson) estimates of the cumulative survivor are given in the last column of Table 6.7. We see that, for these data, a one-year period is short enough for the approximation to be good.   □

6.5   Exercises

(1) The table below gives a series indicating if patients were arriving (indicated by 1) at the intensive care unit of a hospital in the Oxford, England, Regional Hospital Board each day from 4 February 1963 to 18 March 1964 (Lindsey, 1992, p. 26, from Cox and Lewis, 1966, pp. 254–255; read across rows).

00010	00100	10000	10101	10001	00110	10001	01000
00111	00101	01000	10100	10001	00111	00011	00000
01000	01100	00101	10001	01101	01110	11110	01010
10101	00001	01100	10100	11011	11011	01000	00111
01100	00001	10110	01010	01110	00100	01010	00001
01001	00000	01010	01011	01101	01101	00101	10011
00111	00101	00011	00000	11011	00100	01110	01111
11011	00111	11001	11011	01111	10101	11011	11111
00111	11100	10010	11011	10011	10110	10111	00110
00111	00001	11000	11000	01111	00111	10001	01010
00110	00000	1					

(a) Plot the cumulative number of events against time and interpret the resulting graph.
(b) Fit a model to determine if there is any relationship between patients arriving on successive days.
(c) Group the data by month and plot them to see if there is evidence of stationarity.
(d) What can be concluded about any systematic change over time? Consider both steady changes and seasonal effects.

(2) Exercise (2.12) gave the traffic violations each year among male subjects in a driver education study. Develop a Markov chain model to describe these data.

(3) Beveridge (1936) gives the average rates paid to agricultural labourers for threshing and winnowing one rased quarter each of wheat, barley, and oats in each decade from 1250 to 1459. These are payments for performing the manual labour of a given task, not daily wages. He obtained them from the rolls of eight Winchester Bishopric Manors (Downton, Ecchinswel, Overton, Meon, Witney, Wargrave, Wycombe, Farnham) in the south of England. As well, he gives the average daily wages of carpenters and masons in Taunton manor, and the average price of wheat for all England, as shown below.

Agriculture	Carpenter	Mason	Wheat price
3.30	3.01	2.91	4.95
3.37	3.08	2.95	4.52
3.45	3.00	3.23	6.23
3.62	3.04	3.11	5.00
3.57	3.05	3.30	6.39
3.85	3.14	2.93	5.68
4.05	3.12	3.13	7.91
4.62	3.03	3.27	6.79
4.92	2.91	3.10	5.17
5.03	2.94	2.89	4.79
5.18	3.47	3.80	6.96
6.10	3.96	4.13	7.98
7.00	4.02	4.04	6.67
7.22	3.98	4.00	5.17
7.23	4.01	4.00	5.45
7.31	4.06	4.29	6.39
7.35	4.08	4.30	5.84
7.34	4.11	4.31	5.54
7.30	4.51	4.75	7.34
7.33	5.13	5.15	4.86
7.25	4.27	5.26	6.01

(a) Fit an autoregression model to each series.
(b) Compare the results with those for time trend models.
(c) Does the price of wheat display any relationship to the rates paid to agricultural labourers?
(d) Fit a multiple regression model for wheat prices containing an autoregression, a time trend, and a dependence on agricultural rates.
   i. Are all these variables necessary in the model?
   ii. Interpret the results.
   iii. Plot profile likelihood functions for all important parameters.

(4) Table 6.4 gave the evolution of wage increases for low grade jobs; the table below gives a similar series of annual percentage increases in average wages of white collar workers in high grade jobs in Britain, 1962–1979 (Nichols, 1983).

1962	3.5	1971	6.2
1963	3.7	1972	5.6
1964	3.5	1973	5.7
1965	4.2	1974	6.2
1966	4.2	1975	8.8
1967	4.1	1976	6.5
1968	4.7	1977	7.7
1969	5.9	1978	8.8
1970	6.4	1979	8.0

    (a) Plot and compare the two series.

    (b) Find a reasonable model for this series.

    (c) Compare it with the results given for the other series.

    (d) Redo the analyses for the two tables using a log normal distribution.

(5) The judgements of three different levels of albedos by four observers at a level of illumination of 2 were given in Table 6.5. The experiment was also performed, with the same observers, at an illumination of 1.2. The results are given below (McNemar, 1954, p. 321).

	Level of albedos		
Observer	0.07	0.14	0.26
1	11	24	60
2	22	26	44
3	16	22	55
4	20	32	82

    (a) Set up the analysis of variance table.

    (b) Calculate the intra-class correlation and the differences in evaluation with level of albedos.

    (c) Compare your results with those for the lower level of illumination.

    (d) In Section 5.3.2, we performed an analysis of variance with two explanatory variables. Could this be extended to study level of albedos and level of illumination simultaneously?

(6) The Panel Study of Income Dynamics carried out in the USA contains information on unemployment periods due to layoffs. The sample distinguishes two ways in which the unemployment spell could end: by being recalled to the same job or finding a new job. The results are given in the table on the following page (Han and Hausman, 1990). Data for which durations can end in more than one way are called 'competing risks'. Usually, strong assumptions have to be made in order to model them.

Week	New job	Recall	Censor	Week	New job	Recall	Censor
1	10	93	0	36	2	1	0
2	8	118	0	37	0	1	2
3	8	55	0	38	1	0	0
4	23	58	0	39	5	4	7
5	3	18	0	40	4	1	1
6	11	26	0	41	1	0	0
7	1	6	0	42	0	0	2
8	22	38	0	43	1	4	2
9	6	13	1	44	0	0	0
10	7	10	0	45	1	0	0
11	4	4	0	46	0	0	0
12	13	32	1	47	0	0	2
13	10	19	9	48	0	0	1
14	0	9	2	49	1	0	1
15	4	14	2	50	1	1	0
16	10	9	3	51	0	0	0
17	8	7	18	52	4	0	23
18	5	2	6	53	1	0	0
19	2	0	3	54	0	0	0
20	9	12	4	55	0	0	2
21	3	1	7	56	1	0	0
22	5	7	9	57	0	0	1
23	1	0	2	58	0	0	0
24	7	10	4	59	0	0	0
25	2	1	2	60	1	0	1
26	18	15	21	61	0	0	2
27	0	2	1	62	0	0	0
28	0	2	0	63	0	0	0
29	1	0	1	64	0	0	0
30	9	4	9	65	0	0	1
31	0	0	3	66	1	0	1
32	1	0	1	67	0	1	1
33	1	0	0	68	0	0	0
34	2	1	3	69	0	1	0
35	2	0	8	70	4	3	33

(a) Ignoring the reason for unemployment ending, plot the Kaplan–Meier survivor curve.

(b) One possible approach to modelling competing risks is to assume that all terminations except that currently of interest are forms of censoring.

    i. Plot the Kaplan–Meier curve for obtaining a new job, assuming that those recalled are censored (as well as those actually censored).

   ii. Does any simple logistic model fit these data well?

   iii. Use the same approach for recalls, assuming that finding a new job is censoring.

  (c) Discuss possible drawbacks of such an approach to competing risks.

(7) Table 6.6 gave the survival over a ten-year period of women with Stage II cancer of the cervix.

  (a) Construct the life table for these data using both the Kaplan–Meier (binomial) and Aalen–Nelson (Poisson) methods.

  (b) Women with Stage II cancer have a more advanced form of the disease than those with Stage I in Table 6.6. Do the forms of their two survivor curves support this fact?

  (c) Compare the probabilities of censoring in the two stages.

(8) Table 4.18 gave the lengths of marriage before divorce in Liège. Notice that there is no censoring in these data.

  (a) Plot the Kaplan–Meier curve for these data.

  (b) Reconstruct the data as a contingency table and compare the fits of any appropriate logistic models.

  (c) Discuss the complications in interpreting such a graph and models, given the special way in which the data were collected.

7
Where to now?

Doing statistics is exciting. In the analysis of any new data set, surprises, both good and bad, will always occur. The results can never be fully predicted in advance—or the study would not have been carried out in the first place. No two studies, or the resulting data sets, are ever really similar, even in the same subject area, so that there is rarely a dull moment. This should be true for all users of statistics. Many of the readers of this book will become specialists in some field, only using statistics as a tool.

Statistics should aid you throughout any empirical study you do:

(1) showing how to design the study in order to obtain the maximum precise and accurate information most efficiently, at minimum cost;

(2) helping to record observations clearly and effectively;

(3) detecting errors and unlikely values, whether typos, an unsuspected sick calf in an experiment (calves do not talk), or unexpected values leading to scientific discovery;

(4) describing relationships among the variables under study;

(5) suggesting ways in which the data may have been generated;

(6) interpreting the meaning of the results;

(7) communicating the results to others unfamiliar with statistical methods.

For many readers, what you have learnt from the material in this book will serve primarily to enable you to communicate with a trained statistician in an intelligent way. Most of the steps just described will be carried out in consultation with such a person. Unfortunately, statistical training, even in social, medical, and biostatistics, is dominated by mathematicians. The consulting statisticians with whom you must cooperate will generally have little concrete knowledge of the scientific method. They will never, themselves, have been responsible for carrying through a full laboratory experiment or for interviewing a complete sample of people. Communication will not be made easier by this fact.

Other readers of this material may go on to become professional statisticians. You will have a second major reason to find statistics exciting. Similar statistical methods are applicable in a wide variety of subject areas. Practising statistics means encountering specialists in many of these fields and learning many new things with each new project. Again, rarely a dull moment. As well, statistics itself is constantly interacting with these diverse subject matters and being forced to develop new methods to meet their demands. You will probably do so yourself. During your training, seize any chance to spend a year or two in a scientific lab or social survey unit. In your future role as consulting statistician, this time will be invaluable, perhaps more so than all your mathematical statistics courses.

However, all the time spent doing statistical work will certainly not be exciting. Much tedious, although not dull, time must be passed preparing the data, 'cleaning' them, becoming familiar with them by creating enumerable repetitive tables of summary statistics. Much sweat goes into producing a first rate analysis of any data set. If you have worked through the examples and the exercises using a hand calculator, you will already know this. This time has not been lost. It has permitted you to become familiar with some interesting, if simple, data sets from a variety of different fields and to see what is really going on when one does statistics.

Fortunately, a great deal of this tediousness can be alleviated by the use of computers. In everyday practice, it is now rare to carry out analyses without one. After completing the material in this book, it is extremely unlikely that you will ever again have to perform statistical calculations only with a calculator. Many sophisticated statistical packages are available on most types of computers. However, one of the major problems with them is that one rarely knows *exactly* what they are doing when one asks them to make some analysis of data. Different packages will often provide rather different results when asked to do the same thing. With your knowledge of what statistical analyses are really about, you should be able to check that your package is doing what you really want from it.

Up until fairly recently, statistical packages on computers were primarily useful for the normal linear regression and analysis of variance models, covered in Chapter 5, as well as standard summary statistics, such as means and variances. Now any self-respecting statistical package should have reliable routines for doing the log linear and logistic models of Chapter 2 and, most likely, options to handle some of the models, such as autoregression or life tables, from Chapter 6, as well as good graphical facilities. Thus, at present, minimum requirements if you are planning to acquire a statistical package are

- direct input of data tables from standard data files;
- manipulation of large data sets;
- means of modifying and performing calculations on observations;
- production of summary or descriptive statistics;
- creation of high resolution graphics, such as histograms and scatterplots;
- multiple linear regression and analysis of variance, hopefully, using the same procedures for both;
- log linear and logistic models;
- time series models, such as autoregression;
- life table models, often called survival models.

More sophisticated packages will also include

- an integrated means of fitting all members of the family of generalised linear models in a similar way;
- three-dimensional graphics;
- dynamically animating graphics, such as rotation and brushing;
- simulation methods;
- ways of programming your own specially conceived models;
- specialised optimisation routines;
- a journal recording everything (except perhaps the graphics) that has happened during a session at the computer (essential for teaching).

Unfortunately, little software is yet available to handle the fitting of a wide variety of probability distributions, as we did in Chapter 4. Statistical packages that will allow fitting regression models for this choice of distributions or a range of general models for dependent response data, to which a brief introduction was given in Chapter 6, are only slowly beginning to appear.

If you have gone through the material in this book using a computer to work out the examples and exercises, I do hope that you have also looked closely at the explanations of how the actual calculations are carried out. Remember too that the software that you have struggled to master is unlikely to be that which you will use in your subsequent professional work.

The usual approach to introducing statistics concentrates on hypothesis testing, linear regression, and analysis of variance. This leads the majority of students to wonder of what use it all is. In contrast, in this book, you have encountered the basic *principles* of statistical *modelling*. However, you have only had a small taste of the vast number of models available, and constantly being developed, and of their power in elucidating the world around you.

You should now have some idea of what to do when faced with a new study— what strategy to use. The basic steps were outlined at the beginning of this chapter. *All are important.* However, in this book, I have concentrated, perhaps too much, on the modelling stage, the steps (4) to (6) given above. Now is the time to summarise these in a bit more detail. Modelling a new data set involves

(1) making a reasonable choice of one or more possible distributions for the response variable of interest;

(2) considering what kinds of dependence there may be among responses, where appropriate;

(3) selecting appropriate explanatory variables, and checking if they actually do explain changes in the distribution among subgroups in the population;

(4) examining the goodness-of-fit of the resulting model, often through the use of graphical methods and residuals—these will often indicate unusual or unexpected aspects of the data that need to be taken into account in improving the model;

(5) if the first model proves unsatisfactory, repeating the steps with other more appropriate ones;

(6) studying what the finally selected model can tell you about the data generating mechanism;

(7) perhaps, again going back and trying alternative models that rival that selected to see what they can say about the data generating mechanism—remember that *no model is ever true*; it is only a crude, if useful, approximation to the data generating mechanism.

With computer power increasing ever more rapidly, such a sequence of steps can very quickly be performed, even on large data sets. One may be tempted to try every imaginable model and take the best, because this is easily feasible. Some statistical packages even encourage it, with automatic stepwise procedures. It is better to use a calculator and have the time to think about things! Computers make possible a rational modelling strategy, allowing the rapid fitting of a series of models to a data set. But these should be models that make sense in the scientific context in which you are working. Never forget that, in the modelling approach to statistics, what you are looking for are clues to the data generating mechanism underlying your study, not a highly significant P-value.

A key characteristic of a good statistician is to be one who asks new and unusual questions—to the person performing the study (yourself if that is the case) and to the data so acquired. Perhaps this book has given you some idea of how statistics can help in answering some of those questions!

Appendix A
Tables

A.1 *P*-values from the χ^2 distribution

d.f.	0.20	0.10	0.05	0.02	0.01	0.001
			P-value			
1	1.642	2.706	3.841	5.412	6.635	10.83
2	3.219	4.605	5.991	7.824	9.210	13.82
3	4.642	6.251	7.815	9.837	11.345	16.27
4	5.989	7.779	9.488	11.668	13.277	18.47
5	7.289	9.236	11.070	13.388	15.086	20.52
6	8.558	10.645	12.592	15.033	16.812	22.46
7	9.803	12.017	14.067	16.622	18.475	24.32
8	11.030	13.362	15.507	18.168	20.090	26.12
9	12.242	14.684	16.919	19.679	21.666	27.88
10	13.442	15.987	18.307	21.161	23.209	29.59
11	14.631	17.275	19.675	22.618	24.725	31.26
12	15.812	18.549	21.026	24.054	26.217	32.91
13	16.985	19.812	22.362	25.472	27.688	34.53
14	18.151	21.064	23.685	26.873	29.141	36.12
15	19.311	22.307	24.996	28.259	30.578	37.70
16	20.465	23.542	26.296	29.633	32.000	39.25
17	21.615	24.769	27.587	30.995	33.409	40.79
18	22.760	25.989	28.869	32.346	34.805	42.31
19	23.900	27.204	30.144	33.687	36.191	43.82
20	25.038	28.412	31.410	35.020	37.566	45.31
21	26.171	29.615	32.671	36.343	38.932	46.80
22	27.301	30.813	33.924	37.660	40.289	48.27
23	28.429	32.007	35.172	38.968	41.638	49.73
24	29.553	33.196	36.415	40.270	42.980	51.18
25	30.675	34.382	37.652	41.566	44.314	52.62

d.f.	P-value					
	0.20	0.10	0.05	0.02	0.01	0.001
26	31.795	35.563	38.885	42.856	45.642	54.05
27	32.912	36.741	40.113	44.140	46.963	55.48
28	34.027	37.916	41.337	45.419	48.278	56.89
29	35.139	39.087	42.557	46.693	49.588	58.30
30	36.250	40.256	43.773	47.962	50.892	59.70
31	37.359	41.422	44.985	49.226	52.191	61.10
32	38.466	42.585	46.194	50.487	53.486	62.49
33	39.572	43.745	47.400	51.743	54.776	63.87
34	40.676	44.903	48.602	52.995	56.061	65.25
35	41.778	46.059	49.802	54.244	57.342	66.62
36	42.879	47.212	50.998	55.489	58.619	67.99
37	43.978	48.363	52.192	56.730	59.893	69.35
38	45.076	49.513	53.384	57.969	61.162	70.70
39	46.173	50.660	54.572	59.204	62.428	72.05
40	47.269	51.805	55.758	60.436	63.691	73.40
41	48.363	52.949	56.942	61.665	64.950	74.74
42	49.456	54.090	58.124	62.892	66.206	76.08
43	50.548	55.230	59.304	64.116	67.459	77.42
44	51.639	56.369	60.481	65.337	68.710	78.75
45	52.729	57.505	61.656	66.555	69.957	80.08
46	53.818	58.641	62.830	67.771	71.201	81.40
47	54.906	59.774	64.001	68.985	72.443	82.72
48	55.993	60.907	65.171	70.197	73.683	84.04
49	57.079	62.038	66.339	71.406	74.919	85.35
50	58.164	63.167	67.505	72.613	76.154	86.66

A.2 *P*-values from the Student *t* distribution

			P-value			
			One-sided test			
	0.10	0.05	0.025	0.01	0.005	0.0005
			Two-sided test			
d.f.	0.20	0.10	0.05	0.02	0.01	0.001
1	3.078	6.314	12.706	31.821	63.657	636.619
2	1.886	2.920	4.303	6.965	9.925	31.599
3	1.638	2.353	3.182	4.541	5.841	12.924
4	1.533	2.132	2.776	3.747	4.604	8.610
5	1.476	2.015	2.571	3.365	4.032	6.869
6	1.440	1.943	2.447	3.143	3.707	5.959
7	1.415	1.895	2.365	2.998	3.499	5.408
8	1.397	1.860	2.306	2.896	3.355	5.041
9	1.383	1.833	2.262	2.821	3.250	4.781
10	1.372	1.812	2.228	2.764	3.169	4.587
11	1.363	1.796	2.201	2.718	3.106	4.437
12	1.356	1.782	2.179	2.681	3.055	4.318
13	1.350	1.771	2.160	2.650	3.012	4.221
14	1.345	1.761	2.145	2.624	2.977	4.140
15	1.341	1.753	2.131	2.602	2.947	4.073
16	1.337	1.746	2.120	2.583	2.921	4.015
17	1.333	1.740	2.110	2.567	2.898	3.965
18	1.330	1.734	2.101	2.552	2.878	3.922
19	1.328	1.729	2.093	2.539	2.861	3.883
20	1.325	1.725	2.086	2.528	2.845	3.850
21	1.323	1.721	2.080	2.518	2.831	3.819
22	1.321	1.717	2.074	2.508	2.819	3.792
23	1.319	1.714	2.069	2.500	2.807	3.768
24	1.318	1.711	2.064	2.492	2.797	3.745
25	1.316	1.708	2.060	2.485	2.787	3.725
26	1.315	1.706	2.056	2.479	2.779	3.707
27	1.314	1.703	2.052	2.473	2.771	3.690
28	1.313	1.701	2.048	2.467	2.763	3.674
29	1.311	1.699	2.045	2.462	2.756	3.659
30	1.310	1.697	2.042	2.457	2.750	3.646
40	1.303	1.684	2.021	2.423	2.704	3.551
50	1.299	1.676	2.009	2.403	2.678	3.496
60	1.296	1.671	2.000	2.390	2.660	3.460
70	1.294	1.667	1.994	2.381	2.648	3.435
80	1.292	1.664	1.990	2.374	2.639	3.416
90	1.291	1.662	1.987	2.368	2.632	3.402
100	1.290	1.660	1.984	2.364	2.626	3.390

A.3 P-values from the F distribution

Numerator degree of freedom = 1

Denominator	P-value					
d.f.	0.20	0.10	0.05	0.02	0.01	0.001
1	9.47	39.86	161.45	1012.55	4052.18	405284.06
2	3.56	8.53	18.51	48.51	98.50	998.50
3	2.68	5.54	10.13	20.62	34.12	167.03
4	2.35	4.54	7.71	14.04	21.20	74.14
5	2.18	4.06	6.61	11.32	16.26	47.18
6	2.07	3.78	5.99	9.88	13.75	35.51
7	2.00	3.59	5.59	8.99	12.25	29.25
8	1.95	3.46	5.32	8.39	11.26	25.41
9	1.91	3.36	5.12	7.96	10.56	22.86
10	1.88	3.29	4.96	7.64	10.04	21.04
11	1.86	3.23	4.84	7.39	9.65	19.69
12	1.84	3.18	4.75	7.19	9.33	18.64
13	1.82	3.14	4.67	7.02	9.07	17.82
14	1.81	3.10	4.60	6.89	8.86	17.14
15	1.80	3.07	4.54	6.77	8.68	16.59
16	1.79	3.05	4.49	6.67	8.53	16.12
17	1.78	3.03	4.45	6.59	8.40	15.72
18	1.77	3.01	4.41	6.51	8.29	15.38
19	1.76	2.99	4.38	6.45	8.18	15.08
20	1.76	2.97	4.35	6.39	8.10	14.82
21	1.75	2.96	4.32	6.34	8.02	14.59
22	1.75	2.95	4.30	6.29	7.95	14.38
23	1.74	2.94	4.28	6.25	7.88	14.20
24	1.74	2.93	4.26	6.21	7.82	14.03
25	1.73	2.92	4.24	6.18	7.77	13.88
26	1.73	2.91	4.23	6.14	7.72	13.74
27	1.73	2.90	4.21	6.11	7.68	13.61
28	1.72	2.89	4.20	6.09	7.64	13.50
29	1.72	2.89	4.18	6.06	7.60	13.39
30	1.72	2.88	4.17	6.04	7.56	13.29
40	1.70	2.84	4.08	5.87	7.31	12.61
50	1.69	2.81	4.03	5.78	7.17	12.22
60	1.68	2.79	4.00	5.71	7.08	11.97
70	1.67	2.78	3.98	5.67	7.01	11.80
80	1.67	2.77	3.96	5.64	6.96	11.67
90	1.67	2.76	3.95	5.61	6.93	11.57
100	1.66	2.76	3.94	5.59	6.90	11.50
110	1.66	2.75	3.93	5.57	6.87	11.43
120	1.66	2.75	3.92	5.56	6.85	11.38

Numerator degrees of freedom = 2

Denominator	*P*-value					
d.f.	0.20	0.10	0.05	0.02	0.01	0.001
1	12.00	49.50	199.50	1249.50	4999.50	499999.50
2	4.00	9.00	19.00	49.00	99.00	999.00
3	2.89	5.46	9.55	18.86	30.82	148.50
4	2.47	4.32	6.94	12.14	18.00	61.25
5	2.26	3.78	5.79	9.45	13.27	37.12
6	2.13	3.46	5.14	8.05	10.92	27.00
7	2.04	3.26	4.74	7.20	9.55	21.69
8	1.98	3.11	4.46	6.64	8.65	18.49
9	1.93	3.01	4.26	6.23	8.02	16.39
10	1.90	2.92	4.10	5.93	7.56	14.91
11	1.87	2.86	3.98	5.70	7.21	13.81
12	1.85	2.81	3.89	5.52	6.93	12.97
13	1.83	2.76	3.81	5.37	6.70	12.31
14	1.81	2.73	3.74	5.24	6.51	11.78
15	1.80	2.70	3.68	5.14	6.36	11.34
16	1.78	2.67	3.63	5.05	6.23	10.97
17	1.77	2.64	3.59	4.97	6.11	10.66
18	1.76	2.62	3.55	4.90	6.01	10.39
19	1.75	2.61	3.52	4.84	5.93	10.16
20	1.75	2.59	3.49	4.79	5.85	9.95
21	1.74	2.57	3.47	4.74	5.78	9.77
22	1.73	2.56	3.44	4.70	5.72	9.61
23	1.73	2.55	3.42	4.66	5.66	9.47
24	1.72	2.54	3.40	4.63	5.61	9.34
25	1.72	2.53	3.39	4.59	5.57	9.22
26	1.71	2.52	3.37	4.56	5.53	9.12
27	1.71	2.51	3.35	4.54	5.49	9.02
28	1.71	2.50	3.34	4.51	5.45	8.93
29	1.70	2.50	3.33	4.49	5.42	8.85
30	1.70	2.49	3.32	4.47	5.39	8.77
40	1.68	2.44	3.23	4.32	5.18	8.25
50	1.66	2.41	3.18	4.23	5.06	7.96
60	1.65	2.39	3.15	4.18	4.98	7.77
70	1.65	2.38	3.13	4.14	4.92	7.64
80	1.64	2.37	3.11	4.11	4.88	7.54
90	1.64	2.36	3.10	4.09	4.85	7.47
100	1.64	2.36	3.09	4.07	4.82	7.41
110	1.63	2.35	3.08	4.05	4.80	7.36
120	1.63	2.35	3.07	4.04	4.79	7.32

Numerator degrees of freedom = 3

Denominator	P-value					
d.f.	0.20	0.10	0.05	0.02	0.01	0.001
1	13.06	53.59	215.71	1350.50	5403.35	540379.19
2	4.16	9.16	19.16	49.17	99.17	999.17
3	2.94	5.39	9.28	18.11	29.46	141.11
4	2.48	4.19	6.59	11.34	16.69	56.18
5	2.25	3.62	5.41	8.67	12.06	33.20
6	2.11	3.29	4.76	7.29	9.78	23.70
7	2.02	3.07	4.35	6.45	8.45	18.77
8	1.95	2.92	4.07	5.90	7.59	15.83
9	1.90	2.81	3.86	5.51	6.99	13.90
10	1.86	2.73	3.71	5.22	6.55	12.55
11	1.83	2.66	3.59	4.99	6.22	11.56
12	1.80	2.61	3.49	4.81	5.95	10.80
13	1.78	2.56	3.41	4.67	5.74	10.21
14	1.76	2.52	3.34	4.55	5.56	9.73
15	1.75	2.49	3.29	4.45	5.42	9.34
16	1.74	2.46	3.24	4.36	5.29	9.01
17	1.72	2.44	3.20	4.29	5.18	8.73
18	1.71	2.42	3.16	4.22	5.09	8.49
19	1.70	2.40	3.13	4.16	5.01	8.28
20	1.70	2.38	3.10	4.11	4.94	8.10
21	1.69	2.36	3.07	4.07	4.87	7.94
22	1.68	2.35	3.05	4.03	4.82	7.80
23	1.68	2.34	3.03	3.99	4.76	7.67
24	1.67	2.33	3.01	3.96	4.72	7.55
25	1.66	2.32	2.99	3.93	4.68	7.45
26	1.66	2.31	2.98	3.90	4.64	7.36
27	1.66	2.30	2.96	3.87	4.60	7.27
28	1.65	2.29	2.95	3.85	4.57	7.19
29	1.65	2.28	2.93	3.83	4.54	7.12
30	1.64	2.28	2.92	3.81	4.51	7.05
40	1.62	2.23	2.84	3.67	4.31	6.59
50	1.60	2.20	2.79	3.59	4.20	6.34
60	1.60	2.18	2.76	3.53	4.13	6.17
70	1.59	2.16	2.74	3.49	4.07	6.06
80	1.58	2.15	2.72	3.47	4.04	5.97
90	1.58	2.15	2.71	3.45	4.01	5.91
100	1.58	2.14	2.70	3.43	3.98	5.86
110	1.57	2.13	2.69	3.41	3.96	5.82
120	1.57	2.13	2.68	3.40	3.95	5.78

Numerator degrees of freedom = 4

Denominator	P-value					
d.f.	0.20	0.10	0.05	0.02	0.01	0.001
1	13.64	55.83	224.58	1405.83	5624.58	562499.56
2	4.24	9.24	19.25	49.25	99.25	999.25
3	2.96	5.34	9.12	17.69	28.71	137.10
4	2.48	4.11	6.39	10.90	15.98	53.44
5	2.24	3.52	5.19	8.23	11.39	31.09
6	2.09	3.18	4.53	6.86	9.15	21.92
7	1.99	2.96	4.12	6.03	7.85	17.20
8	1.92	2.81	3.84	5.49	7.01	14.39
9	1.87	2.69	3.63	5.10	6.42	12.56
10	1.83	2.61	3.48	4.82	5.99	11.28
11	1.80	2.54	3.36	4.59	5.67	10.35
12	1.77	2.48	3.26	4.42	5.41	9.63
13	1.75	2.43	3.18	4.28	5.21	9.07
14	1.73	2.39	3.11	4.16	5.04	8.62
15	1.71	2.36	3.06	4.06	4.89	8.25
16	1.70	2.33	3.01	3.97	4.77	7.94
17	1.68	2.31	2.96	3.90	4.67	7.68
18	1.67	2.29	2.93	3.84	4.58	7.46
19	1.66	2.27	2.90	3.78	4.50	7.27
20	1.65	2.25	2.87	3.73	4.43	7.10
21	1.65	2.23	2.84	3.69	4.37	6.95
22	1.64	2.22	2.82	3.65	4.31	6.81
23	1.63	2.21	2.80	3.61	4.26	6.70
24	1.63	2.19	2.78	3.58	4.22	6.59
25	1.62	2.18	2.76	3.55	4.18	6.49
26	1.62	2.17	2.74	3.52	4.14	6.41
27	1.61	2.17	2.73	3.50	4.11	6.33
28	1.61	2.16	2.71	3.47	4.07	6.25
29	1.60	2.15	2.70	3.45	4.04	6.19
30	1.60	2.14	2.69	3.43	4.02	6.12
40	1.57	2.09	2.61	3.30	3.83	5.70
50	1.56	2.06	2.56	3.22	3.72	5.46
60	1.55	2.04	2.53	3.16	3.65	5.31
70	1.54	2.03	2.50	3.13	3.60	5.20
80	1.53	2.02	2.49	3.10	3.56	5.12
90	1.53	2.01	2.47	3.08	3.53	5.06
100	1.53	2.00	2.46	3.06	3.51	5.02
110	1.52	2.00	2.45	3.05	3.49	4.98
120	1.52	1.99	2.45	3.04	3.48	4.95

TABLES

Numerator degrees of freedom = 5

Denominator	P-value					
d.f.	0.20	0.10	0.05	0.02	0.01	0.001
1	14.01	57.24	230.16	1440.61	5763.65	576404.56
2	4.28	9.29	19.30	49.30	99.30	999.30
3	2.97	5.31	9.01	17.43	28.24	134.58
4	2.48	4.05	6.26	10.62	15.52	51.71
5	2.23	3.45	5.05	7.95	10.97	29.75
6	2.08	3.11	4.39	6.58	8.75	20.80
7	1.97	2.88	3.97	5.76	7.46	16.21
8	1.90	2.73	3.69	5.22	6.63	13.48
9	1.85	2.61	3.48	4.84	6.06	11.71
10	1.80	2.52	3.33	4.55	5.64	10.48
11	1.77	2.45	3.20	4.34	5.32	9.58
12	1.74	2.39	3.11	4.16	5.06	8.89
13	1.72	2.35	3.03	4.02	4.86	8.35
14	1.70	2.31	2.96	3.90	4.69	7.92
15	1.68	2.27	2.90	3.81	4.56	7.57
16	1.67	2.24	2.85	3.72	4.44	7.27
17	1.65	2.22	2.81	3.65	4.34	7.02
18	1.64	2.20	2.77	3.59	4.25	6.81
19	1.63	2.18	2.74	3.53	4.17	6.62
20	1.62	2.16	2.71	3.48	4.10	6.46
21	1.61	2.14	2.68	3.44	4.04	6.32
22	1.61	2.13	2.66	3.40	3.99	6.19
23	1.60	2.11	2.64	3.36	3.94	6.08
24	1.59	2.10	2.62	3.33	3.90	5.98
25	1.59	2.09	2.60	3.30	3.85	5.89
26	1.58	2.08	2.59	3.28	3.82	5.80
27	1.58	2.07	2.57	3.25	3.78	5.73
28	1.57	2.06	2.56	3.23	3.75	5.66
29	1.57	2.06	2.55	3.21	3.73	5.59
30	1.57	2.05	2.53	3.19	3.70	5.53
40	1.54	2.00	2.45	3.05	3.51	5.13
50	1.52	1.97	2.40	2.97	3.41	4.90
60	1.51	1.95	2.37	2.92	3.34	4.76
70	1.50	1.93	2.35	2.88	3.29	4.66
80	1.50	1.92	2.33	2.86	3.26	4.58
90	1.49	1.91	2.32	2.84	3.23	4.53
100	1.49	1.91	2.31	2.82	3.21	4.48
110	1.49	1.90	2.30	2.81	3.19	4.45
120	1.48	1.90	2.29	2.80	3.17	4.42

Numerator degrees of freedom = 6

Denominator				P-value		
d.f.	0.20	0.10	0.05	0.02	0.01	0.001
1	14.26	58.20	233.99	1464.45	5858.99	585937.12
2	4.32	9.33	19.33	49.33	99.33	999.33
3	2.97	5.28	8.94	17.25	27.91	132.85
4	2.47	4.01	6.16	10.42	15.21	50.53
5	2.22	3.40	4.95	7.76	10.67	28.83
6	2.06	3.05	4.28	6.39	8.47	20.03
7	1.96	2.83	3.87	5.58	7.19	15.52
8	1.88	2.67	3.58	5.04	6.37	12.86
9	1.83	2.55	3.37	4.65	5.80	11.13
10	1.78	2.46	3.22	4.37	5.39	9.93
11	1.75	2.39	3.09	4.15	5.07	9.05
12	1.72	2.33	3.00	3.98	4.82	8.38
13	1.69	2.28	2.92	3.84	4.62	7.86
14	1.67	2.24	2.85	3.72	4.46	7.44
15	1.66	2.21	2.79	3.63	4.32	7.09
16	1.64	2.18	2.74	3.54	4.20	6.80
17	1.63	2.15	2.70	3.47	4.10	6.56
18	1.62	2.13	2.66	3.41	4.01	6.35
19	1.61	2.11	2.63	3.35	3.94	6.18
20	1.60	2.09	2.60	3.30	3.87	6.02
21	1.59	2.08	2.57	3.26	3.81	5.88
22	1.58	2.06	2.55	3.22	3.76	5.76
23	1.57	2.05	2.53	3.19	3.71	5.65
24	1.57	2.04	2.51	3.15	3.67	5.55
25	1.56	2.02	2.49	3.13	3.63	5.46
26	1.56	2.01	2.47	3.10	3.59	5.38
27	1.55	2.00	2.46	3.07	3.56	5.31
28	1.55	2.00	2.45	3.05	3.53	5.24
29	1.54	1.99	2.43	3.03	3.50	5.18
30	1.54	1.98	2.42	3.01	3.47	5.12
40	1.51	1.93	2.34	2.88	3.29	4.73
50	1.49	1.90	2.29	2.80	3.19	4.51
60	1.48	1.87	2.25	2.75	3.12	4.37
70	1.47	1.86	2.23	2.71	3.07	4.28
80	1.47	1.85	2.21	2.68	3.04	4.20
90	1.46	1.84	2.20	2.66	3.01	4.15
100	1.46	1.83	2.19	2.65	2.99	4.11
110	1.46	1.83	2.18	2.63	2.97	4.07
120	1.45	1.82	2.18	2.62	2.96	4.04

Numerator degrees of freedom = 8

Denominator	P-value					
d.f.	0.20	0.10	0.05	0.02	0.01	0.001
1	14.58	59.44	238.88	1494.99	5981.07	598144.12
2	4.36	9.37	19.37	49.37	99.37	999.37
3	2.98	5.25	8.85	17.01	27.49	130.62
4	2.47	3.95	6.04	10.16	14.80	49.00
5	2.20	3.34	4.82	7.50	10.29	27.65
6	2.04	2.98	4.15	6.14	8.10	19.03
7	1.93	2.75	3.73	5.33	6.84	14.63
8	1.86	2.59	3.44	4.79	6.03	12.05
9	1.80	2.47	3.23	4.41	5.47	10.37
10	1.75	2.38	3.07	4.13	5.06	9.20
11	1.72	2.30	2.95	3.91	4.74	8.35
12	1.69	2.24	2.85	3.74	4.50	7.71
13	1.66	2.20	2.77	3.60	4.30	7.21
14	1.64	2.15	2.70	3.48	4.14	6.80
15	1.62	2.12	2.64	3.39	4.00	6.47
16	1.61	2.09	2.59	3.30	3.89	6.19
17	1.59	2.06	2.55	3.23	3.79	5.96
18	1.58	2.04	2.51	3.17	3.71	5.76
19	1.57	2.02	2.48	3.12	3.63	5.59
20	1.56	2.00	2.45	3.07	3.56	5.44
21	1.55	1.98	2.42	3.02	3.51	5.31
22	1.54	1.97	2.40	2.99	3.45	5.19
23	1.53	1.95	2.37	2.95	3.41	5.09
24	1.53	1.94	2.36	2.92	3.36	4.99
25	1.52	1.93	2.34	2.89	3.32	4.91
26	1.52	1.92	2.32	2.86	3.29	4.83
27	1.51	1.91	2.31	2.84	3.26	4.76
28	1.51	1.90	2.29	2.82	3.23	4.69
29	1.50	1.89	2.28	2.80	3.20	4.64
30	1.50	1.88	2.27	2.78	3.17	4.58
40	1.47	1.83	2.18	2.64	2.99	4.21
50	1.45	1.80	2.13	2.56	2.89	4.00
60	1.44	1.77	2.10	2.51	2.82	3.86
70	1.43	1.76	2.07	2.48	2.78	3.77
80	1.42	1.75	2.06	2.45	2.74	3.70
90	1.42	1.74	2.04	2.43	2.72	3.65
100	1.41	1.73	2.03	2.41	2.69	3.61
110	1.41	1.73	2.02	2.40	2.68	3.58
120	1.41	1.72	2.02	2.39	2.66	3.55

			Numerator degrees of freedom = 12			
Denominator			*P*-value			
d.f.	0.20	0.10	0.05	0.02	0.01	0.001
1	14.90	60.71	243.91	1526.31	6106.32	610667.81
2	4.40	9.41	19.41	49.42	99.42	999.42
3	2.98	5.22	8.74	16.76	27.05	128.32
4	2.46	3.90	5.91	9.89	14.37	47.41
5	2.18	3.27	4.68	7.23	9.89	26.42
6	2.02	2.90	4.00	5.88	7.72	17.99
7	1.91	2.67	3.57	5.06	6.47	13.71
8	1.83	2.50	3.28	4.53	5.67	11.19
9	1.76	2.38	3.07	4.15	5.11	9.57
10	1.72	2.28	2.91	3.87	4.71	8.45
11	1.68	2.21	2.79	3.65	4.40	7.63
12	1.65	2.15	2.69	3.48	4.16	7.00
13	1.62	2.10	2.60	3.34	3.96	6.52
14	1.60	2.05	2.53	3.23	3.80	6.13
15	1.58	2.02	2.48	3.13	3.67	5.81
16	1.56	1.99	2.42	3.05	3.55	5.55
17	1.55	1.96	2.38	2.97	3.46	5.32
18	1.53	1.93	2.34	2.91	3.37	5.13
19	1.52	1.91	2.31	2.86	3.30	4.97
20	1.51	1.89	2.28	2.81	3.23	4.82
21	1.50	1.87	2.25	2.76	3.17	4.70
22	1.49	1.86	2.23	2.73	3.12	4.58
23	1.49	1.84	2.20	2.69	3.07	4.48
24	1.48	1.83	2.18	2.66	3.03	4.39
25	1.47	1.82	2.16	2.63	2.99	4.31
26	1.47	1.81	2.15	2.60	2.96	4.24
27	1.46	1.80	2.13	2.58	2.93	4.17
28	1.46	1.79	2.12	2.56	2.90	4.11
29	1.45	1.78	2.10	2.54	2.87	4.05
30	1.45	1.77	2.09	2.52	2.84	4.00
40	1.41	1.71	2.00	2.38	2.66	3.64
50	1.39	1.68	1.95	2.30	2.56	3.44
60	1.38	1.66	1.92	2.25	2.50	3.32
70	1.37	1.64	1.89	2.21	2.45	3.23
80	1.37	1.63	1.88	2.19	2.42	3.16
90	1.36	1.62	1.86	2.17	2.39	3.11
100	1.36	1.61	1.85	2.15	2.37	3.07
110	1.35	1.61	1.84	2.14	2.35	3.04
120	1.35	1.60	1.83	2.12	2.34	3.02

A.4 Area under the standard normal curve

$|z_i|$	0.00	0.01	0.02	0.03	0.04
0.0	0.0000	0.0040	0.0080	0.0120	0.0160
0.1	0.0398	0.0438	0.0478	0.0517	0.0557
0.2	0.0793	0.0832	0.0871	0.0910	0.0948
0.3	0.1179	0.1217	0.1255	0.1293	0.1331
0.4	0.1554	0.1591	0.1628	0.1664	0.1700
0.5	0.1915	0.1950	0.1985	0.2019	0.2054
0.6	0.2257	0.2291	0.2324	0.2357	0.2389
0.7	0.2580	0.2611	0.2642	0.2673	0.2704
0.8	0.2881	0.2910	0.2939	0.2967	0.2995
0.9	0.3159	0.3186	0.3212	0.3238	0.3264
1.0	0.3413	0.3438	0.3461	0.3485	0.3508
1.1	0.3643	0.3665	0.3686	0.3708	0.3729
1.2	0.3849	0.3869	0.3888	0.3907	0.3925
1.3	0.4032	0.4049	0.4066	0.4082	0.4099
1.4	0.4192	0.4207	0.4222	0.4236	0.4251
1.5	0.4332	0.4345	0.4357	0.4370	0.4382
1.6	0.4452	0.4463	0.4474	0.4484	0.4495
1.7	0.4554	0.4564	0.4573	0.4582	0.4591
1.8	0.4641	0.4649	0.4656	0.4664	0.4671
1.9	0.4713	0.4719	0.4726	0.4732	0.4738
2.0	0.4772	0.4778	0.4783	0.4788	0.4793
2.1	0.4821	0.4826	0.4830	0.4834	0.4838
2.2	0.4861	0.4864	0.4868	0.4871	0.4875
2.3	0.4893	0.4896	0.4898	0.4901	0.4904
2.4	0.4918	0.4920	0.4922	0.4925	0.4927
2.5	0.4938	0.4940	0.4941	0.4943	0.4945
2.6	0.4953	0.4955	0.4956	0.4957	0.4959
2.7	0.4965	0.4966	0.4967	0.4968	0.4969
2.8	0.4974	0.4975	0.4976	0.4977	0.4977
2.9	0.4981	0.4982	0.4982	0.4983	0.4984
3.0	0.4987	0.4987	0.4987	0.4988	0.4988
3.1	0.4990	0.4991	0.4991	0.4991	0.4992
3.2	0.4993	0.4993	0.4994	0.4994	0.4994
3.3	0.4995	0.4995	0.4995	0.4996	0.4996
3.4	0.4997	0.4997	0.4997	0.4997	0.4997

$|z_i|$	0.05	0.06	0.07	0.08	0.09
0.0	0.0199	0.0239	0.0279	0.0319	0.0359
0.1	0.0596	0.0636	0.0675	0.0714	0.0753
0.2	0.0987	0.1026	0.1064	0.1103	0.1141
0.3	0.1368	0.1406	0.1443	0.1480	0.1517
0.4	0.1736	0.1772	0.1808	0.1844	0.1879
0.5	0.2088	0.2123	0.2157	0.2190	0.2224
0.6	0.2422	0.2454	0.2486	0.2517	0.2549
0.7	0.2734	0.2764	0.2794	0.2823	0.2852
0.8	0.3023	0.3051	0.3078	0.3106	0.3133
0.9	0.3289	0.3315	0.3340	0.3365	0.3389
1.0	0.3531	0.3554	0.3577	0.3599	0.3621
1.1	0.3749	0.3770	0.3790	0.3810	0.3830
1.2	0.3944	0.3962	0.3980	0.3997	0.4015
1.3	0.4115	0.4131	0.4147	0.4162	0.4177
1.4	0.4265	0.4279	0.4292	0.4306	0.4319
1.5	0.4394	0.4406	0.4418	0.4429	0.4441
1.6	0.4505	0.4515	0.4525	0.4535	0.4545
1.7	0.4599	0.4608	0.4616	0.4625	0.4633
1.8	0.4678	0.4686	0.4693	0.4699	0.4706
1.9	0.4744	0.4750	0.4756	0.4761	0.4767
2.0	0.4798	0.4803	0.4808	0.4812	0.4817
2.1	0.4842	0.4846	0.4850	0.4854	0.4857
2.2	0.4878	0.4881	0.4884	0.4887	0.4890
2.3	0.4906	0.4909	0.4911	0.4913	0.4916
2.4	0.4929	0.4931	0.4932	0.4934	0.4936
2.5	0.4946	0.4948	0.4949	0.4951	0.4952
2.6	0.4960	0.4961	0.4962	0.4963	0.4964
2.7	0.4970	0.4971	0.4972	0.4973	0.4974
2.8	0.4978	0.4979	0.4979	0.4980	0.4981
2.9	0.4984	0.4985	0.4985	0.4986	0.4986
3.0	0.4989	0.4989	0.4989	0.4990	0.4990
3.1	0.4992	0.4992	0.4992	0.4993	0.4993
3.2	0.4994	0.4994	0.4995	0.4995	0.4995
3.3	0.4996	0.4996	0.4996	0.4996	0.4997
3.4	0.4997	0.4997	0.4997	0.4997	0.4998

A.5 Values of the gamma function

To calculate other values, use $\Gamma(a+1) = a\Gamma(a)$.

a	$\Gamma(a)$	a	$\Gamma(a)$	a	$\Gamma(a)$	a	$\Gamma(a)$
1.00	1.0000	1.25	0.9064	1.50	0.8862	1.75	0.9191
1.01	0.9943	1.26	0.9044	1.51	0.8866	1.76	0.9214
1.02	0.9888	1.27	0.9025	1.52	0.8870	1.77	0.9238
1.03	0.9835	1.28	0.9007	1.53	0.8876	1.78	0.9262
1.04	0.9784	1.29	0.8990	1.54	0.8882	1.79	0.9288
1.05	0.9735	1.30	0.8975	1.55	0.8889	1.80	0.9314
1.06	0.9687	1.31	0.8960	1.56	0.8896	1.81	0.9341
1.07	0.9642	1.32	0.8946	1.57	0.8905	1.82	0.9368
1.08	0.9597	1.33	0.8934	1.58	0.8914	1.83	0.9397
1.09	0.9555	1.34	0.8922	1.59	0.8924	1.84	0.9426
1.10	0.9514	1.35	0.8912	1.60	0.8935	1.85	0.9456
1.11	0.9474	1.36	0.8902	1.61	0.8947	1.86	0.9487
1.12	0.9436	1.37	0.8893	1.62	0.8959	1.87	0.9518
1.13	0.9399	1.38	0.8885	1.63	0.8972	1.88	0.9551
1.14	0.9364	1.39	0.8879	1.64	0.8986	1.89	0.9584
1.15	0.9330	1.40	0.8873	1.65	0.9001	1.90	0.9618
1.16	0.9298	1.41	0.8868	1.66	0.9017	1.91	0.9652
1.17	0.9267	1.42	0.8864	1.67	0.9033	1.92	0.9688
1.18	0.9237	1.43	0.8860	1.68	0.9050	1.93	0.9724
1.19	0.9209	1.44	0.8858	1.69	0.9068	1.94	0.9761
1.20	0.9182	1.45	0.8857	1.70	0.9086	1.95	0.9799
1.21	0.9156	1.46	0.8856	1.71	0.9106	1.96	0.9837
1.22	0.9131	1.47	0.8856	1.72	0.9126	1.97	0.9877
1.23	0.9108	1.48	0.8857	1.73	0.9147	1.98	0.9917
1.24	0.9085	1.49	0.8859	1.74	0.9168	1.99	0.9958
						2.00	1.0000

A.6   Estimating the Weibull power parameter

$\frac{\bar{y}^2}{s^2+\bar{y}^2}$	$\hat{\alpha}$	$\frac{\bar{y}^2}{s^2+\bar{y}^2}$	$\hat{\alpha}$	$\frac{\bar{y}^2}{s^2+\bar{y}^2}$	$\hat{\alpha}$	$\frac{\bar{y}^2}{s^2+\bar{y}^2}$	$\hat{\alpha}$
0.0000	0.05	0.6243	1.30	0.8498	2.55	0.9205	3.80
0.0000	0.10	0.6408	1.35	0.8542	2.60	0.9222	3.85
0.0000	0.15	0.6562	1.40	0.8584	2.65	0.9239	3.90
0.0040	0.20	0.6708	1.45	0.8624	2.70	0.9255	3.95
0.0143	0.25	0.6845	1.50	0.8663	2.75	0.9270	4.00
0.0331	0.30	0.6973	1.55	0.8700	2.80	0.9285	4.05
0.0596	0.35	0.7095	1.60	0.8735	2.85	0.9300	4.10
0.0920	0.40	0.7209	1.65	0.8769	2.90	0.9314	4.15
0.1283	0.45	0.7317	1.70	0.8802	2.95	0.9328	4.20
0.1667	0.50	0.7420	1.75	0.8833	3.00	0.9341	4.25
0.2057	0.55	0.7516	1.80	0.8863	3.05	0.9354	4.30
0.2445	0.60	0.7607	1.85	0.8892	3.10	0.9367	4.35
0.2822	0.65	0.7694	1.90	0.8920	3.15	0.9379	4.40
0.3186	0.70	0.7776	1.95	0.8947	3.20	0.9391	4.45
0.3533	0.75	0.7854	2.00	0.8973	3.25	0.9402	4.50
0.3863	0.80	0.7928	2.05	0.8998	3.30	0.9413	4.55
0.4174	0.85	0.7998	2.10	0.9022	3.35	0.9424	4.60
0.4467	0.90	0.8065	2.15	0.9046	3.40	0.9435	4.65
0.4742	0.95	0.8129	2.20	0.9068	3.45	0.9445	4.70
0.5000	1.00	0.8189	2.25	0.9090	3.50	0.9455	4.75
0.5242	1.05	0.8247	2.30	0.9111	3.55	0.9465	4.80
0.5469	1.10	0.8302	2.35	0.9131	3.60	0.9474	4.85
0.5682	1.15	0.8354	2.40	0.9150	3.65	0.9484	4.90
0.5881	1.20	0.8404	2.45	0.9169	3.70	0.9493	4.95
0.6068	1.25	0.8452	2.50	0.9187	3.75	0.9502	5.00

Bibliography

1. Agresti, A. (1984) *Analysis of Ordinal Categorical Data*. New York: John Wiley.
2. *Agresti, A. (1990) *Categorical Data Analysis*. New York: John Wiley.
3. Aitkin, M.A., Francis, B., Hinde, J., and Anderson, D. (1989) *Statistical Modelling in GLIM*. Oxford: Oxford University Press.
4. Aiuppa, T.A. (1988) Evaluation of Pearson curves as an approximation of the maximum probable annual aggregate loss. *Journal of Risk and Insurance* **55**, 425–441.
5. Blalock, H.M. (1972) *Social Statistics*. New York: McGraw-Hill.
6. Brass, W. (1959) Simplified methods of fitting the truncated negative binomial distribution. *Biometrika* **45**, 59–68.
7. *Breslow, N.E. and Day, N.E. (1982) *Statistical Methods in Cancer Research*. Volume 1. *The Analysis of Case-control studies*. Lyon: International Agency for Research on Cancer.
8. Burridge, J. (1981) Empirical Bayes analysis of survival time data. *Journal of the Royal Statistical Society* **B43**, 65–75.
9. Chatfield, C., Ehrenberg, A.S.C., and Goodhardt, G.J. (1966) Progress on a simplified model of stationary purchasing behaviour. *Journal of the Royal Statistical Society* **B28**, 317–367
10. *Clayton, D. and Hills, M. (1993) *Statistical Models in Epidemiology*. Oxford: Oxford University Press.
11. *Collett, D. (2002) *Modelling Binary Data*, 2nd edn. London: Chapman and Hall.
12. Cox, D.R. and Lewis, P.A.W. 1966, *The Statistical Analysis of Series of Events*. London: Methuen, pp. 254–255.
13. Dahiya, R.C. and Gross, A.J. (1973) Estimating the zero class from a truncated Poisson sample. *Journal of the American Statistical Association* **68**, 731–733.
14. Davis, C.S. (2002) *Statistical Methods for the Analysis of Repeated Measurements*. Berlin: Springer-Verlag.
15. *Derman, C., Gleser, L.J., and Olkin, I. (1973) *A Guide to Probability Theory and Application*. New York: Holt, Rinehart, and Winston.
16. Desai, A. (1973) *Water Facilities for the Untouchables in Rural Gujarat*. Delhi: ICSSR.
17. *Dobson, A.J. (1990) *An Introduction to Generalized Linear Models*. London: Chapman and Hall.
18. Erickson, B.H. and Nosanchuk, T.A. (1977) *Understanding Data*. Toronto: McGraw-Hill Ryerson.
19. *Fienberg, S.E. (1977) *The Analysis of Cross-Classified Categorical Data*. Cambridge: MIT Press.

20. *Fingleton, B. (1984) *Models of Category Counts.* Cambridge: Cambridge University Press.
21. Fisher, R.A. (1938) *Sankhyā* **4**, 14–17.
22. Fisher, R.A. (1958) *Statistical Methods for Research Workers.* Edinburgh: Oliver and Boyd.
23. Fox, W.R. and Lasker, G.W. (1983) The distribution of surname frequencies. *International Statistical Review* **51**, 81–87.
24. Geissler, A. (1889) Beiträge zur Frage des Geschlechtsverhältnisses der Geborenen. *Zeitschrift des Königl. Sächsischen Statistischen Bureaus* **35**, 1–24.
25. Gelfand, A.E. and Dalal, S.R. (1990) A note on overdispersed exponential families. *Biometrika* **77**, 55–64
26. Gelman, A. and Nolan, D. (2002) *Teaching Statistics. A Bag of Tricks.* Oxford: Oxford University Press.
27. Graubard, B.I. and Korn, E.L. (1987) Choice of column scores for testing independence in ordered $2 \times K$ contingency tables. *Biometrics* **43**, 471–476.
28. Haberman, S.J. (1974) Log-linear models for frequency tables with ordered classifications. *Biometrics* **30**, 589–600.
29. *Haberman, S.J. (1978) *Analysis of Qualitative Data. Volume 1. Introductory Topics.* New York: Academic Press.
30. Han, A. and Hausman, J.A. (1990) Flexible parametric estimation of duration and competing risk models. *Journal of Applied Econometrics* **5**, 1–28.
31. *Hosmer, D.W. and Lemeshow, S. (1989) *Applied Logistic Regression,* New York: John Wiley.
32. Irwin, J.O. (1975) The generalized Waring distribution. Part II. *Journal of the Royal Statistical Society* **A138**, 204–284.
33. Jarrett, R.G. (1979) A note on the intervals between coal-mining disasters. *Biometrika* **66**, 191–193.
34. *Kalbfleisch, J.G. (1985) *Probability and Statistical Inference. I. Probability. II. Statistical Inference.* Berlin: Springer-Verlag.
35. Lancaster, T. (1972) A stochastic model for the duration of a strike. *Journal of the Royal Statistical Society* **A135**, 257–271.
36. Lazarsfeld, P. (1955) The interpretation of statistical relations as a research operation. In Lazarsfeld and Rosenberg, *The Language of Social Research.* pp. 115–125. New York: Free Press.
37. Lindsey, J.K. (1978) *Primary Education in Bombay: Introduction to a Social Study.* Oxford: Pergamon.
38. Lindsey, J.K. (1981) Social class in the educational system: an international comparison. *Canadian Review of Sociology and Anthropology* **18**, 299–319.
39. Lindsey, J.K. (1989) *The Analysis of Categorical Data Using GLIM.* Berlin: Springer-Verlag.
40. Lindsey, J.K. (1992) *The Analysis of Stochastic Processes Using GLIM.* Berlin: Springer-Verlag.
41. *Lindsey, J.K. (1995) *Modelling Frequency and Count Data.* Oxford: Oxford University Press.

42. Lindsey, J.K. and Mersch, M. (1992) Fitting and comparing probability distributions with log linear models. *Computational Statistics and Data Analysis* **13**, 373–384.
43. Lombard, H.L. and Doering, C.R. (1947) Treatment of the fourfold table by partial association and partial correlation as it relates to public health problems. *Biometrics* **3**, 123–128.
44. Maguire, B.A., Pearson, E.S., and Wynn, A.S.A. (1952) The time intervals between industrial accidents. *Biometrika* **39**, 168–180.
45. McNemar, Q. (1954) *Psychological Statistics.* New York: John Wiley.
46. McPherson, G. (1990) *Statistics in Scientific Investigation. Its Basis, Application, and Interpretation.* Berlin: Springer-Verlag.
47. Mueller, J.H., Schuessler, K.F., and Costner, H.L. (1970) *Statistical Reasoning in Sociology.* Boston: Houghton Mifflin.
48. Nelson, J.F. (1980) Multiple victimisation in American cities: a statistical analysis of rare events. *American Journal of Sociology* **85**, 870–891.
49. Nichols, D.A. (1983) Macroeconomic determinants of wage adjustments in white-collar occupations. *Review of Economics and Statistics* **65**, 203–213.
50. Skellam, J.G. (1948) A probability distribution derived from the binomial distribution by regarding the probability of success as variable between sets of trials. *Journal of the Royal Statistical Society* **B10**, 257–261.
51. Sokal, R.R. and Rohlf, F.J. (1969) *Biometry. The Principles and Practice of Statistics in Biological Research.* San Francisco: W. H. Freeman.
52. Beveridge, W. (1936) Wages in the Winchester manors. *Economic History Review* **7**, 22–43.
53. Yule, G.U. and Kendall, M.G. (1950) *An Introduction to the Theory of Statistics.* London: Griffin.
54. Zelterman, D. (1987) Goodness-of-fit tests for large sparse multinomial distributions. *Journal of the American Statistical Association* **82**, 624–629.

Author index

Subject index

Breinigsville, PA USA
23 December 2010
252076BV00002B/2/P